MODELING AND SIMULATION FOR ANALYZING GLOBAL EVENTS

MODELING AND SIMULATION FOR ANALYZING GLOBAL EVENTS

John A. Sokolowski

Catherine M. Banks

WILEY

A JOHN WILEY & SONS, INC., PUBLICATION

Cover art: Whitney A. Sokolowski.

Published by John Wiley & Sons, Inc. Hoboken, New Jersey.
Published simultaneously in Canada.

For general information on our other products and services or for technical support, please contact
our Customer Care Department within the United States at (800) 762-2974, outside the United
States at (317) 572-3993 or fax (317) 572-4002.

Wiley also publishes its books in a variety of electronic formats. Some content that appears in print
may not be available in electronic books. For more information about Wiley products, visit our
web site at www.wiley.com.

Library of Congress Cataloging-in-Publication Data:

Sokolowski, John A., 1953–
  Modeling and simulation for analyzing global events / John A. Sokolowski, Catherine M.
Banks.
        p. cm.
  Includes bibliographical references and index.
  ISBN 978-0-470-47841-7 (cloth)
  1. Event history analysis. 2. Event history analysis–Computer simulation. 3. History–
Mathematical models. 4. Social sciences–Mathematical models. I. Banks, Catherine M., 1960–
II. Title.
  H61.S7757 2009
  001.4—dc22

                                                                        2008053470

10 9 8 7 6 5 4 3 2 1

This book is dedicated to

my grandmother, Theresa Hanz
who taught me what is important in life

—John A. Sokolowski

my family

—Catherine M. Banks

CONTENTS

PREFACE

The impetus for this book was the development of an academic course that integrates qualitative analysis of real-world events into quantitative, numerical representations of those data to produce models that enhance traditional social science modeling. As we explored textbooks and sources to support this course, we quickly learned that none existed. Hence, we determined that a text in modeling and simulation with applications to global events would serve not only our students' needs, but the needs of many engineering, science, and social science majors who want to explore various methods of researching and representing the international events that shape the world in which we live.

Modeling and simulation (M&S) as its own discipline has a number of modeling domains in its toolbox. Students with backgrounds in M&S are no doubt familiar with system dynamics, game theory, agent-based modeling, and social network modeling. Students with liberal arts backgrounds, such as in international studies, political science, history, and sociology, are accustomed to traditional social science methods of modeling, such as statistical modeling, formal modeling (game theory), and agent-based modeling. It did not take long for us to recognize that coupling these disciplines and their methods of investigation would facilitate an interdisciplinary approach to discovery and communication using the common ground between social scientists and engineers and scientists: modeling and simulation.

It is important to note, however, that engineering students approach modeling and simulation from a very different perspective than do students of international studies, political science, and sociology. Engineering majors are able to draw on their mathematical background and engineering training for model development, and from their computer science skills for visualization (communication). Social science students engage in vigorous qualitative assessment as they explore "what happened." Understandably, social science students model those data without the skill-set of the engineer and the sophistication of an engineering model. The engineering student, on the other hand, may not fully appreciate the complexity of human behavior and social network analysis as uncovered in the qualitative research conducted by a student trained in the liberal arts. Hence, students with social science backgrounds would be well served to acquire skills that enable them to better analyze events that shape our world. Students with engineering and science backgrounds may already possess the skills necessary to model, simulate, and visualize; however, they would benefit by applying those skills to real-world events that affect our lives.

Many universities have realized that modeling and simulation has become an important tool in understanding and even solving numerous and diverse problems. These institutions have begun to offer introductory and application courses in this field to acquaint their students with the foundational concepts that will help them employ modeling and simulation in their various disciplines. This book provides an orientation to the theory and applications of modeling and simulation in the realm of social science, with an emphasis on international studies, political science, history, and sociology. It is designed to provide an understanding of historical and contemporary global events and to represent those events in a scientific format. Attention is focused on (1) understanding the event with an eye to gathering empirical data to construct a model, an abstract representation of the event; (2) understanding how to construct a model; (3) understanding what model would best facilitate the representation of the data; and (4) understanding how to analyze, verify, and validate the model.

The text is appropriate for students of all disciplines, but will be of special interest to students observing international affairs because it introduces global events that all of us should understand. The intent is twofold: (1) to expose students to events in political history that have placed us in the environment in which we now live, and (2) to encourage students to explore other ways to understand and discuss that environment.

To students we offer a concise look at the key concepts in the field of modeling and simulation. Although modeling and simulation necessarily entails mathematical representations and computer programs, the authors endeavored to highlight these concepts so that non-engineering and non-science students could understand the modeling and simulation concepts covered in the book.

The book is divided into three parts with nine chapters. In Part I, *Principles of Modeling and Simulation: Advancing Global Studies*, we introduce modeling and simulation and its role in research and analysis. In Chapter 1, *Modeling and Simulation: What, When, and Why*, we provide a brief history of modeling and simulation, list the many uses or applications of modeling and simulation, and speak to the advantages and disadvantages of using models in problem solving. Chapter 2 focuses on *Research*

Methodologies for Modeling Global Events. We discuss the social sciences as a discipline, the differences between qualitative and quantitative research, a methodology for applying qualitative and quantitative research, and how modeling and simulation can be used specifically in analyzing global events.

In Part II, *Modeling Paradigms*, we examine four methods of modeling. In Chapter 3, *System Dynamics*, we explain a fundamental approach that allows us to investigate complex structures and institutions. Chapter 4, *Agent-Based Modeling and Social Networks*, illustrates the dynamism of this type of modeling because the goal of the research is to imitate actions and interactions among the agents as units of analysis. The section on social networks elaborates on a type of modeling that helps us to understand the connections among people, whether they be political leaders, specific groups, and/or cliques in organizations. In Chapter 5, *Game Theory*, we discuss this type of modeling as a tool to study the interactions of individuals, also called agents, in various contexts, such as social dilemmas or combative situations.

In Part III, *Modeling Global Events*, we introduce four real-world events as case studies to which the four modeling paradigms can be applied. The case studies are also representative of fundamental areas of study for social science students: a look at internal commotion within an anarchic state; a multilayered study of the Solidarity movement in Poland as part of the last days of Soviet communism, including representations of the people, leaders, governments (ideologies), militaries, and church and state; a unilateral military intervention; and the issue of compellence and deterrence during a national security crisis.

PART I
Principles of Modeling and Simulation: Advancing Global Studies

1 Modeling and Simulation: What, When, and Why

INTRODUCTION

Modeling and simulation (M&S) is becoming an academic program of choice for students in all disciplines.[1] It is a discipline with its own body of knowledge, theory, and research methodology at whose core is the fundamental notion that *models are approximations of the real-world*. First a *model* is created approximating an event. The model is then followed by *simulation*, which allows for repeated observation of the model. After one or many simulations of the model, the third step, *analysis*, takes place. Analysis aids in the ability to draw conclusions, verify and validate research, and make recommendations based on various iterations or simulations of the model. These basic precepts, coupled with *visualization*, the ability to represent data as a way to interface with the model, make M&S a problem-based discipline that allows for repeated testing of a hypothesis. Teaching these precepts and providing research and development opportunities are the core to M&S education. M&S also serves as a tool or application that expands the ability to analyze and communicate new research or findings.

Our intent with this book is to introduce some of the modeling applications used in M&S education and research with a focus on global events so that students interested in international studies and related social sciences can acquire the skills necessary to this critical technology. Fundamental

[1]This chapter is based largely on Catherine M. Banks, What is modeling and simulation, in *Principles of Modeling and Simulation: A Multidisciplinary Approach*, by John A. Sokolowski and Catherine M. Banks, eds. New York: John Wiley & Sons, Inc., 2009.

Modeling and Simulation for Analyzing Global Events, By John A. Sokolowski and Catherine M. Banks
Copyright © 2009 John Wiley & Sons, Inc.

to this critical technology is an understanding of the four precepts that define the discipline of M&S: modeling, simulation, analysis, and visualization. Students study the basics of *modeling* as a way to understand the various modeling paradigms appropriate for conducting digital computer simulations. They must understand *simulation* and the methodology, development, verification and validation, and design of simulation experiments. Students who are able to engage in *visualization* are able to provide an overview of interactive, real-time, three-dimensional computer graphics and visual simulations using high-level development tools. Important to all student research is an *analysis* of the findings, and included in any good analysis is an observation of the constraints and requirements in using M&S. In other words, analysis includes making known the limitations of the research.

It was political scientist Herbert A. Simon (1916–2001) who introduced the notion of *learning by doing* (also known as *experiential learning*).[2] M&S can be just that. It is the simulation of a model that allows for imitation of the operation of a real-world process or system over time. To do this, one must generate a history, real or artificial, to draw inferences concerning the operating characteristics of the real system that is represented [1]. The art and science of M&S has evolved very rapidly since the mid-1980s—so much so that it easily parallels the technological advances of mainframe and desktop computers and the ever-increasing emergence of the Internet and World Wide Web (www).

AN OVERVIEW OF MODELING AND SIMULATION

A **model** is a representation of events and/or things that are real (a case study) or contrived (a use-case). It can be a representation of an actual system, or it can be something used in lieu of the real thing to better understand a certain aspect of that thing. To produce a model, one must abstract from reality a description of a vibrant system. The model can depict the system at some point of abstraction or at multiple levels

[2]Herbert A. Simon was a political scientist who conducted research in a variety of disciplines, including cognitive psychology, computer science, public administration, economics, management, and philosophy of science. Simon was among the founding fathers of several of today's most important scientific domains, including artificial intelligence, information processing, decision making, problem solving, organization theory, complex systems, and computer simulation of scientific discovery. He was the first person to analyze the architecture of complexity and to propose a preferential attachment mechanism to explain power law distributions. He introduced the notions of experiential learning and bounded rationality. Simon's research at Carnegie Mellon University resulted in numerous publications. He was one of the most influential social scientists of the twentieth century.

of the abstraction, with the goal of representing the system in a mathematically reliable fashion. A **simulation** is an applied methodology that can describe the behavior of that system using either a mathematical or a symbolic model [2]. Simply, simulation is the imitation of the operation of a real-world process or system over a period of time [1]. In this book we introduce the various uses of simulation in the analysis of global events. For example, simulation can be used to represent the effect of changes in governmental policy during a counterinsurgency, to analyze the decision-making processes of opposing military leaders, or to assess the social network structure of a political leader and his/her circle of advisers.

M&S begins with (1) the development of a computer simulation or design based on a model of an actual or theoretical physical system, then (2) execution of that model on a digital computer, and (3) analysis of the output. Modeling and the ability to act-out with those models provide a credible way to understand the complexity and particulars of a real entity [2]. From these three steps one can see that M&S facilitates the simulation of a system such as a social network structure and then the testing of a hypothesis related to that structure. For example, if an analyst wanted to determine how intelligence (or the lack of) affects the decisions of an international aid agency supervisor in a dismal and unsafe theater of operations, he could develop a social network structure that includes representations for the supervisor, the "intel" agency, and the aid workers who respond to the determined distribution–assistance operation. This will facilitate measuring the outcome of the plan.

It is important to note that *models are driven by data*, so the data collection must be done with great accuracy. To develop this model the analyst needs to research the aid intervention strategy (case study) by answering who, what, where, when, why, and how. The analyst should note the supervisor's experience, goals, expectations, and actions. Then he or she should assess the intelligence provided and the supervisor's response to the intel (resulting ultimately in the supervisor's decision or action). Execution of the supervisor's orders affects another component in the social network structure, the aid workers. Their experience, goals, expectations, and actions should also be reviewed. Finally, the results of the execution should be assessed. Assessment of the outcome (something the supervisor would also do) should include two questions: (1) *Were the plan and execution done right?* (2) *Was this the right thing to do?* The analyst is tasked with two primary responsibilities: carrying out the research required to develop the model and assessing the outcome.

Once a model is created, the analyst can craft a fairly well thought out and credible hypothesis that delves more deeply into the case study. For example, *if the supervisor was given this additional piece of intelligence, the following might have been the result*. But even that needs to be weighed carefully. There may be unexpected changes in the model—say that the additional intel indicated that all was clear for an aid distribution operation but it did not take into consideration indiscriminate skirmishes between factions in the region. What can the supervisor do to accommodate an unpredictable occurrence of violence, one not necessarily aimed at the aid operation? The analysts can assist the supervisor by creating a number of simulations or iterations of the model to ascertain the "what if." Upon reviewing the output of the simulations, the supervisor can review the data and determine an alternative plan of action should an outbreak of violence occur.

From the example above one can see that M&S affords analysts the ability to repeat the testing of a hypothesis through various simulations. Simulations are very important. Unlike the terms *model* and *modeling*, defining simulation is not as clear-cut. Definitions of *simulation* cover a wide range:

- A method of implementing a model over time
- A technique for testing, analysis, or training in which real-world systems are used or where real-world and conceptual systems are reproduced by a model
- An unobtrusive scientific method of inquiry involving experiments with a model rather than with the portion of reality that the model represents
- A methodology for extracting information from a model by observing the behavior of the model as it is executed
- A nontechnical term meaning *not real*, *imitation* (the correct word here is the adjective *simulated*)

Simulation is used when a real system cannot be engaged. This may happen when the real system (1) might not be accessible, (2) it might be dangerous to engage the system, (3) it might be unacceptable to engage the system, or (4) the system might simply not exist. To counter these objections, a computer will *imitate* operations of the various real-world facilities or processes. Most modeling depends on computational science for the visualization and simulation of complex, large-scale phenomena. These models may be used to replicate complex systems that exhibit chaotic behavior; thus, simulation must be used to provide a more detailed view of the system. Simulation also allows for virtual reality research whereby the analyst is immersed within the simulated world through the use of devices such as head-mounted display, data gloves, freedom sensors, and forced-feedback elements [2].

Understanding what comprises a model and what constitutes a simulation leads to the next step, which is to couple this with visualization. M&S coupled with visualization refers to developing a model of a system, extracting information from the model (simulation) and using visualization to enhance understanding or interpretation of that information. Thus far we have mentioned a *model of a system* a number of times but have not defined a system.

A **system** is a construct or collection of different elements that together produce results not obtainable using the elements alone.[3] The elements can include people, hardware, software, facilities, policies, and documents: all things required to produce system-level qualities, properties, characteristics, functions, behavior, and performance. Importantly, the value of the system as a whole is the relationship among the parts. There are two types of systems: (1) *discrete*, in which the variables change instantaneously at separate points in time, and (2) *continuous*, where the state variables change continuously with respect to time. There are a number of ways to study a system [3]:

- The actual system versus a model of the system
- A physical versus a mathematical representation
- Analytical solution versus simulation solution (which exercises the simulation for inputs in question to see how they affect the output measures of performance)

It is becoming widely accepted that M&S holds a significant place in research and development, due to its inherent properties of modeling, simulating, analyzing, and visualizing (communicating). Many people in the M&S community (researchers, academicians, and persons in the industry and the military) were introduced to M&S as a tool that evolved with the military of the twentieth century. But its origins can be traced to an ancient military organization whose use of wargames made it one of the most efficient armies in military history.

---

[3]Additional information and definitions can be found in the INCOSE online glossary at http://www.incose.org/mediarelations/glossaryofseterms.aspx.

A BRIEF HISTORY OF MODELING AND SIMULATION

The act of wargames and challenging or outwitting an opponent on the battlefield is centuries old. In Ancient Rome the then world's largest empire was secured by the world's largest military. The Roman army conducted live training between two contingents of its own military (red team versus blue team). Their training battlefield reflected an environment that the troops would encounter somewhere within the vast Roman Empire, which spanned the Scottish border in northern Europe throughout North Africa and into Central Asia. The Roman army had to learn how to fight in unknown regions against armies with diverse warring techniques. Although their training exercises were not intended to draw blood, their training honed a military prowess that made the Roman army the greatest military the world had known for centuries (circa 500 B.C.E. to 1500 C.E.). Significant as they were in Rome, models were not restricted to the art of wargames and military training.

During the Renaissance (1200 to 1600 C.E.), the rebirth of classical learning, artists and scientists were using models in their designs of statuary and edifices. The models were presented to the artist's patron or commissioner as a way of seeking approval of a design before beginning an expensive project such as a marble bust or sarcophagus. One of the most notable artist–scientists of the time was Leonardo DaVinci. He is famous for his paintings, sculptures, building designs, and scientific experiments. His projects include the design of advanced weaponry (including tanks and submarines), flying machines, municipal construction, canals, and ornamental architecture (churches and fortresses) as well as his well-known anatomical studies. Among his many assignments as a military engineer, DaVinci designed a bridge to span the Golden Horn (a freshwater waterway dividing the city) in Constantinople (modern-day Istanbul). DaVinci was also commissioned to do a life-sized equestrian statue (which was later redesigned to be four times larger). To do this he studied the movement of horses, made countless sketches, and devised new casting techniques. He was not able to complete the project, but he had succeeded in making a 22-foot clay model. This artist–engineer made repeated uses of modeling to test the design of many of his inventions and projects. He determined that by understanding how each separate machine part functioned, he could modify it and combine it with other parts in different ways to improve existing machines or to create new machines. He provided one of the first systematic explanations of how machines work and how the elements of machines can be combined. Around this time, a new competition was being introduced in Europe. It came in the form of a game that required intellect and prowess—chess.

The current game of chess as most Westerners know it had its origins in southern Europe in the second half of the fifteenth century. That game was a derivation of a seventh-century game of Indian origin. Included on the chessboard are a king, queen, bishop, knight, rook, and pawn. The object of the game is to checkmate the opponent's king by placing the king under immediate duress, or *check*, with maneuvering such that there is no way for the opponent to remove his king from attack. Think about what is created on the chessboard: a simulated battlefield with two armies that possess equal strength of force. It is now up to the human commander (the chess player) to conduct his hypotheses: *What if I move this way? What will happen if I do this? How will my opponent respond? What is he planning?* The ultimate *checkmate* is rewarded by winning the war (the game). But what if one's opponent is a computer? Can it outwit a human opponent? Yes, it can.

In 1997, an IBM chess-playing computer named Deep Blue won a short six-game exhibition match (not a world title match) by two wins to one, with three draws, against the Russian world champion, Garry Kasparov, after he made a blunder in the opening of the last game. Kasparov accused Deep Blue (IBM) of cheating and demanded a rematch, but officials at IBM declined. His accusation stemmed from the fact that he saw deep intelligence and creativity in the machine's moves, suggesting that during the second game human chess players intervened, in violation of the rules. IBM's response was that the human intervention occurred only between games as the rules provided, so that the developers could modify the program between games. This gave IBM an opportunity to modify weaknesses in the computer's play as it was displayed during the game. Doing these modifications precluded the computer from falling into previous traps set by Kasparov [4]. There are a number of theoreticians who have developed extensive chess strategies and tactics. Many who play the game cite chess as one of the first *board* wargames. By the eighteenth century, military modeling, simulation, and training took on a new perspective.

In the 1780s, with England at the height of its naval power, a Scotsman named John Clerk developed a method of using model ships to gain tactical insights. He used his ships to step through battles analyzing the influence that the geometry of the combatants had on their combat power. Although a military simulation, Clerk's work was not considered a wargame because it did not provide a way to measure or apply the effects of actions—the reward and risk from game theory [5]. On the European continent, however, wargames were being developed formally by the Prussians (modern-day northeastern Germany).

Prussia attained its greatest importance in the eighteenth and nineteenth centuries. In the eighteenth century, during the reign of the Soldier King, Frederick I (1713–1740), Prussia instituted a *standing army*, an army composed of full-time professional soldiers who *stand over* and never disband, even during times of peace. As a result of this significant military capacity, Prussia became a great European power during the latter half of the century under the reign of Frederick II (1740–1786). The Prussians saw the advantages of playing wargames, and by 1824 games were incorporated in training throughout the Prussian army. During the nineteenth century Prime Minister Otto von Bismarck pursued a policy of uniting the German principalities into a *Lesser Germany*, excluding the Austrian Empire. This led to the unification of Germany in 1871. Wargaming contributed to the outstanding military capability of Prussia's standing army and its success on the battlefield during the nineteenth century.

In the United States, Major W. R. Livermore of the Army Corps of Engineers introduced modern wargaming to the American military [5]. In 1883 he translated the German rules to a wargame they had developed based on the American Civil War and Prussia's own wars of 1866 and 1870–1871. When comparing the German attrition tables to actual statistics, Livermore found that errors had been made. He determined that the German attrition tables usually predicted lower casualties than the historical record indicated and chose to adjust his tables accordingly. Upon improving the wargame with the historically accurate data, Livermore sought official acceptance of wargaming for the U.S. military. Much to his surprise, he was blocked by General William T. Sherman, who was serving as the U.S. Army's chief of staff. Sherman felt that wargames depicted soldiers as blocks of wood rather than as human beings. He therefore refused to integrate wargames into military training. Four years after Sherman's refusal to use wargames, the Naval War College decided that it would use Livermore's model. In 1887 the college introduced its first Army–Navy field exercise. By the turn of the twentieth century, wargames made their way into U.S. military exercises and training, but these games lacked the capability and the complexity to model an event with the accuracy now seen in military modeling, a change that came about with the introduction of technology.

The *Link flight simulator*, patented in 1929 by an American, Edward Link, was a pilot trainer resembling a toy airplane, with short wooden wings and a fuselage mounted on a universal joint. Link used organ bellows, driven by an electric pump, to make the trainer pitch and roll as the pilot worked the controls. The Link simulator was used in both the government and private sectors [6]. In 1931 the simulator was fully instrumented and sold to the U.S. Navy. The U.S. Army took delivery of Link trainers three years later. The simulators provided great economy for the military, as vast sums of money and time were saved in the training of Navy and Army pilots in simulators replicating air flight. This is a good example of how using simulation allows the military or any other company or organization to test a system before investing in a full-scale model and to train a person in a less expensive environment.

In the post–World War I period, both the Navy and the Marine Corps employed wargames as part of their training. This training proved useful with the coming of World War II. Under the leadership of General George C. Marshall, live simulation was introduced into military training. As a result, military M&S made quick inroads into training the military of a new world power. With the end of the two world wars, a new period of military engagement was beginning, one that brought with it weapons of mass destruction that required computer-assisted air defense systems to interfere with their delivery. This post–World War II period was known as the *Cold War*. It took place between the two leading world powers, the United States and the Soviet Union, and it lasted almost half a century (1945–1989) as military competition between the two.

On August 29, 1949 the Soviet Union detonated an atomic device at the Semipalatinsk Test Site in Kazakhstan, making it the second nation in the world to detonate a nuclear mechanism. This action served as the impetus for the U.S. government to give grave consideration to the threat of another nation possessing nuclear military capability. As a result, the Department of Defense was given the approval to invest research funds in air defense systems. By the winter of 1949, digital computers were engaged in creating simulated combat. Developed by the U.S. Air Force, a semiautomated ground environment—SAGE—simulated combat from the perspective of more than one combatant. This type of simulation provided military training that now incorporated an air-defense system.

By the 1950s, computers were being used to generate model behavior followed by simulation programs. These computers were then required to process the results of the elements of the simulation-based problem-solving environments [7]. Digital radar was now able to transmit from the newly developed microwave early warning (MEW) radar.

This innovative research was being conducted by engineers at the Massachusetts Institute of Technology (MIT). Significant to the research was a transmission that tied the MEW located at Hanscom Field to the digital computer named Whirlwind located at MIT in Cambridge. Also at this time a scaled-down version of SAGE was being developed. Dubbed the Cape Cod System, this simulator was introduced in 1953. It allowed radar operators and weapons controllers to react to simulated targets presented to them exactly as these targets would appear in an engagement.

The country was becoming embroiled in a military contest that called for technology far beyond the imagination of the average citizen.

Interestingly, some of that same technology was making its way into the homes of many families in everyday, ordinary appliances and communication devices that brought a new element to the postmodern age. In essence, as the country was developing militarily, so was every other aspect of technology—that is why the 1950s were so progressive. This was a unique time in the social history of the country. It was both an age of innocence and a postmodern world with technical advances that would send human beings into outer space. Ironically, it was the newly invented RCA FlipTop television and Regency TR1 transistor radio that delivered fear and talk of war with the Soviets into the American family living room.

At the close of his two-term presidency, Dwight D. Eisenhower (1953–1961) gave an address to the American people about the effects of the ongoing military competition with the Soviet Union. Eisenhower's *military–industrial complex speech* made Americans aware that *a vital element in keeping the peace is our military establishment*. The President emphasized that U.S. arms must be mighty, ready for instant action, so that no potential aggressor might be tempted to risk its own destruction [8]. To do this the federal government would support and fund research that would make the military state of the art, always ahead of the opposition. The President's speech referred to the increasing military buildup in the United States throughout the 1950s. That buildup fueled the nation's growing economy, and many were living quite comfortably during this time. Perhaps somewhat oblivious to what was truly happening, Eisenhower was compelled to explain to his fellow citizens the ramifications of coupling an immense military establishment with an expanding arms industry. This was a new concept for Americans. In fact, the military–industrial complex was a new American experience which had an economic and political influence that reverberated throughout the country. By 1960 the increased spending for this complex amounted to more than half of the U.S. federal expenditure; and as the complex grew, so did the workforce. From the close of World War II (1945) to the end of Eisenhower's second term (1961), an expansive workforce of civilian employees constituted much of the defense industry. Additionally, many universities thrived on the increased research opportunities.

Throughout the 1960s, military wargames became much more sophisticated, moving from strictly tactical training to strategic commands. Games were now incorporating such things as the political capacity of a state or leader. They also became technically mature. This became apparent with work done at the universities. In 1961 a student at MIT created an interactive computer game called Spacewar [9]. The game required the player to operate his or her spaceship during a conflict that was fought with the firing of torpedoes. Pilots of the spaceships scored points by launching missiles that inflicted damage on the opponent, avoiding direct hits by the opponent, and maneuvering the spaceship to avoid the gravitational pull of the sun. This computer game was one of the first *interactive* games in the country. The President and Congress were also pushing forward a research agenda at government institutions. Just over a decade after Spacewar, two engineers with the National Aeronautics and Space Administration (NASA) at Moffett Field, California developed another computer game, one a bit more complex, called Mazewar [10]. This game was networked, and it introduced the concept of online players as *avatars* (a graphical image of a user or a graphical personification of a computer or a computer process) in the form of an eyeball chasing other players around a maze.[4] Mazewar's development in 1974 served as a catalyst for a number of versions on various programs.

The military was also making contributions to M&S by formalizing simulation as a training tool. In 1971 the Navy's *Top Gun* school opened to train fleet fighter pilots in air combat tactics. In 1975 the Tactical Advanced Combat Direction and Electronic Warfare (TACDEW) simulator was being used for team training. The simulator created twenty-two separate shipboard mock-ups with the ability to generate a *virtual* (or *synthetic*) threat environment [11]. Work was also being done with training simulators. Fighter plane cockpits such as the B-52 (long-range heavy bomber aircraft) were simulated so that they could operate with tanker (refueling aircraft) simulators to facilitate training plane–tanker rendezvous.

By 1983, simulator networking was advancing rapidly. The Defense Advanced Research Projects Agency (DARPA) had initiated simulator networking—SIMNET—with an emphasis on tactical team performance on the battlefield. The U.S. Army supported the idea of incorporating armor, mechanized infantry, helicopters, artillery, communications, and logistics into the model for a much more expansive simulated training experience. Combatants could now see each other and communicate over radios. The SIMNET simulator was introduced at the platoon level in 1986. By 1990 over 250 networked simulators at eleven different sites were delivered to the U.S. Army [12]. It wasn't long before the benefit of SIMNET training was realized.

[4]Note the difference between a game and a simulation. A *game* is concerned with entertaining and there is much more player participation. A *simulation* is focused on getting the details of the model and system correct. A simulation does not require a participant or player; a game does.

On July 25, 1990, Saddam Hussein convened a meeting with the U.S. Ambassador to Iraq, April Glaspie, expressing his contempt for two of his Persian Gulf neighbors, Kuwait and the United Arab Emirates. He specifically accused Kuwait of exceeding the Organization for Petroleum Exporting Countries (OPEC) production limits and thus driving down oil prices. This lowering of prices was having a negative affect on the Iraqi economy and he faulted the United States for encouraging this high level of production. Additionally, his aggressive behavior earlier in the year resulted in the cessation of American aid. No more American aid meant that he would look elsewhere to make up for the shortfall—and that elsewhere was Kuwait. Within two weeks of his meeting with Ambassador Glaspie, Saddam ordered his troops into Kuwait. Iraqi troops entered Kuwait on August 2, and six days later an international coalition that included U.S. ground forces was conducting Operation Desert Shield to counter the Iraqi invasion of Kuwait. By January 1991 the U.S.-led international coalition's mission changed to include offensive air attacks. A seamless transition from Operation Desert Shield to Operation Desert Storm was under way.

In February a decisive tank battle was in progress. The Battle of 73 Easting was fought between armored forces of the U.S. Army and the Iraqi Republican Guard. The U.S. ground unit was outnumbered and outgunned, yet it was able to affect the enemy by destroying 85 tanks, 40 personnel carriers, and 30 wheeled vehicles carrying antiaircraft artillery. Why was this outnumbered and outgunned U.S. armored unit able to conduct itself with such precision and success? The answer is because this unit had trained intensively before the engagement using SIMNET. The 73 Easting Project was a collaborative study conducted jointly by the independent Institute for Defense Analyses (IDA), DARPA, and the U.S. Army. With the development of a database and the use of modern computer simulation technology, the soldiers were able to train in a virtual re-creation of the minute-to-minute activities of each participating tank, armored vehicle, truck, and infantry team.

After the engagement, more information was collected by extensive engineering surveys of the battlefield. Exhaustive participant interviews were included. This information was further integrated into the simulation of the battle for future training. The Battle of 73 Easting and the postanalysis proved the significance of computer simulation training in and of itself and in future training with the ability to test alternative cause-and-effect hypotheses with factual and counterfactual analysis. SIMNET would now include conducting controlled experiments by changing key characteristics of the historical event, then refighting the simulated battle and observing the effects on the presumed outcome.

Tied to the events in Iraq was the establishment of the Executive Council on Modeling and Simulation by the U.S. Department of Defense in 1990. In 1991 a Defense Modeling and Simulation Office was established with large investments to advance modular semiautomated forces—ModSAF—a set of software modules and applications used to construct *advanced distributed simulation* and *computer-generated forces* applications. ModSAF modules and applications allow a single operator to create and control large numbers of entities that are used for realistic training, test, and evaluation on the virtual battlefield. Funding also went into the advancement of the Joint Simulation System (JSIMS) to develop training tools and future M&S capability, in particular simulation improvement.

Advancements in computer software and hardware as well as artificial intelligence and software agents have hastened the pace of the maturation of M&S as a discipline and tool. These additional elements that now comprise M&S enhance the capabilities of simulation for more complex phenomena, such as the human personality, in social and conflictual simulations. In the early 1990s, military M&S practitioners began to explore ways to link stand-alone simulations used to model and represent distinct real-world functions into a federation of simulations where simulation entities were given semiautomated behaviors, commonly called *semiautomated forces*. The initial efforts to link simulations showed promise and led to standards in simulation data exchange and the establishment of protocols for creating simulation federations [7].

As semiautomated behaviors were being explored, a closer look at human behavior was under way. It was during this time that a new type of modeling was taking form: **behavioral modeling**, a model of human activity in which individual or group behaviors are derived from the psychological or social aspects of human beings. Behavioral models include a diversity of approaches; however, the computational approaches to human behavior modeling that are most prevalent are social network models and multiagent systems.[5] Behavioral modeling can be used to provide qualitative analysis of a specific foreign leader, to assess the movement of civilian populations in duress, or to understand how culture and religion can affect social behavior [13]. Behavioral modeling allows for the incorporation of socially dependent aspects of behavior that occur when people are together. This type of modeling is now being used in fields of study that include observations of human behavior, be they individual, group,

---

[5]A *social network model* is a model of social behavior that takes into account relationships derived from statistical analysis of relational data. A multiagent system focuses on the way in which social behavior emerges from the actions of agents [15].

or crowd behavior. The areas of education, psychology, industry, and transportation are just some of those using behavioral modeling.

The Department of Defense has also made use of behavioral modeling. In fact, behavioral modeling research has become a significant component of military M&S. The M&S academic community and the Department of Defense analysis community has now expanded its research to include social network analysis and crowd modeling. The M&S industry stayed close to military applications, however, focusing their work on meeting the needs of the U.S. military's Simulation, Training, and Instrumentation Command (STRICOM) in Orlando, Florida, the Air Force Research Laboratory in Dayton, Ohio, the National Simulation Center, and the U.S. Army Training and Doctrine Command (TRADOC), both at Fort Leavenworth, Kansas. The U.S Joint Forces Command (USJFCOM) in Suffolk, Virginia also became very involved in M&S. In 1997 the USJFCOM partnered with Old Dominion University in establishing the Virginia Modeling, Analysis and Simulation Center (VMASC).[6]

It was not long, however, before the explosive growth of computer games for entertainment and the emergence of new uses for M&S shifted the focus of the industry. By the latter half of the 1990s, companies active in the military M&S industry began exploring a variety of new uses for M&S. Today, M&S can be found in just about every research and training institution or venue. M&S is being used in the medical and health care fields, logistics and transportation, manufacturing and distribution, communications, and virtual reality and gaming applications for both entertainment and education. As you will see, M&S has expanded into many application areas.

WHY USE MODELING AND SIMULATION

Once a primary training mechanism for the military, M&S is now being used in a variety of domains, including medical modeling, emergency management, crowd modeling, transportation, game-based learning, and

[6]The Virginia Modeling, Analysis, and Simulation Center (VMASC) is a multidisciplinary modeling, simulation, and visualization collaborative research center managed through the Office of Research at Old Dominion University. VMASC supports the university's modeling and simulation graduate degree programs, offering multidisciplinary M&S Master's and Ph.D. degrees to students across the colleges of engineering and technology, sciences, social sciences, education, and business. With numerous industry, government, and academic members, VMASC furthers the development and application of modeling, simulation, and visualization as an enterprise decision-making tool and promotes economic development. http://www.vmasc.odu.edu.

engineering design, to name a few. Within various forms of media, M&S has already made inroads into a number of liberal arts disciplines. With the advent of modeled and simulated historic events, television has made significant strides in portraying the ancient world as historians have researched and perceived it to be. Advertisements for complex wireless telephone systems have employed simulated, visualized geographic data that scan large crowds, then zero in on the individual user, all in the name of selling the most extensive wireless phone service. Avatars have replaced humans for interfacing within complex communication systems. These applications of M&S have all evolved in a somewhat indirect fashion, and they are a part of our lives, although we may not recognize or even realize that this is M&S. However, as an academic tool, the focus of this book, M&S has a more formal role.

M&S applications are used primarily for analysis, experimentation, and training. Analysis refers to an investigation of a model's behavior. Experimentation occurs when the behavior of the model changes under conditions that exceed the model's design boundaries. Training is the development of knowledge, skills, and abilities obtained as we operate the system represented by the model. Thus, M&S is multifaceted and can be used as a tool, an enabling technology. It is this property that facilitates the use of M&S in many disciplines. What is becoming more and more apparent to traditional producers and users of M&S is that there is a "richness of the possibilities ... and synergies with related disciplines" [14].

M&S can be applied in any field where experimentation is conducted using dynamic models. This includes all types of engineering and science studies as well as social science, business, medical, and education domains. M&S is often the only tool capable of solving complex problems because it allows for an understanding of system dynamics and includes enabling technology, both of which provide a means to explore credible solutions.

Two types of modeling and simulation activity that can be distinguished, depending on whether or not the simulation program runs independent of the system it represents. **Stand-alone simulation** follows the H. A. Simon notion of *learn by doing*, or train as you operate. **Integrated simulation** is used to enrich and support real systems. For many nonengineering and nonscience students, stand-alone simulation may work best. Often, stand-alone application areas are grouped into five categories [7]:

1. *Training.* The goal of training is to provide real-world experience and opportunities in a controlled environment.

2. *Decision support.* Decision support provides either a descriptive, explanatory, predictive tool or an evaluative, prescriptive tool.
3. *Understanding.* This type of modeling and simulation facilitates testing a hypothesis relative to the structure and function of a complex system.
4. *Education and learning.* These are used in teaching and learning systems with dynamic behavior and with serious gaming (also called *game-based learning*).
5. *Entertainment.* Simulation provides a realistic representation for elements possessing dynamic behavior.

Training, decision support, and understanding aim to provide a level of proficiency. An example of stand-alone simulation for training is a simulation used to create an environment that focuses on game theory. Because it focuses on game theory, the simulation may be a **zero-sum simulation** intent on honing the user's decision-making and communication skills. This is done by enveloping the user in confrontational simulations at different levels, or in peace operations, or in conflict management, and then allowing him or her to negotiate a solution. Another example of stand-alone simulation for training is **virtual simulation**, with limited environmental interactions used to develop the motor skills of the user.

Whether it is training, decision support, understanding, or education and learning, M&S applications can be found in a number of social science research areas. For social scientists the traditional methods of modeling include statistical modeling, formal modeling, and agent-based modeling. **Statistical modeling** is the conventional method for the discovery and interpretation of patterns in large numbers of events. **Formal modeling** is a method that provides a rigorous analytic specification of the choices that actors can make and how those choices interact to produce outcomes. **Agent-based modeling** allows for the observation of aggregate behaviors that emerge from the interactions of large numbers of autonomous actors.

Integrating this traditional modeling and analysis capacity with other forms of modeling (simulation and visualization) serves as a tool for expanding and communicating the social scientist's grasp of the subject area being investigated and for providing a much denser schematic for the engineer's model. This relationship will accelerate interdisciplinary research efforts on the part of engineers and social scientists, and it is a very good response to changing research requirements.

As the need for social science qualitative and quantitative analysis expands into various domains, a number of areas can be explored, such as:

- *A foreign policy issue of sanctions:* a system dynamics approach to measuring the effects of failed sanctions on behavior of the civilian population, the insurgency, and/or the controlling regime
- *A global and national issue of energy dependency:* a study of the economic, political, and social capacity to manage energy crises
- *A national issue of immigration:* an analysis of the layered effects (labor, education, health care) of illegal immigrants in a specific community (state, region, or national)
- *A national education issue of game-based learning:* the creation of educational games in the subjects of history and geography
- *Voting habits in countries with dubious election processes:* development of a model that couples statistical data with empirical findings to address voter participation
- *A local government issue of urban design:* continued research in modeling urban development to prescribe improvements in land-use, transportation, and infrastructure

This list introduces just a few of the potential M&S applications in the social sciences. There are also many advantages to M&S. In 1998 the Institute of Industrial Engineers (IIE) listed these advantages (and disadvantages) of using modeling and simulation [1]. From their list it is easy to see why many would choose to apply M&S to research and training. Here are some of the processes and results of using modeling and simulation:

- *Choosing correctly* by testing every aspect of a proposed change without committing additional resources
- *Compressing and expanding time* to allow the user to speed-up or slow-down behavior or phenomena to facilitate in-depth research
- *Understanding why* by reconstructing and examining the scenario closely by controlling the system
- *Exploring possibilities* in the context of policies, operating procedures, and methods without disrupting the actual or real system
- *Diagnosing problems* by understanding the interactions among variables that comprise complex systems
- *Identifying constraints* by reviewing delays on process, information, and materials to ascertain whether or not the constraint is the effect or the cause
- *Developing understanding* by observing how a system operates rather than predicting how it will operate

- *Visualizing the plan* with the use of animation to observe the system or organization actually operating
- *Building consensus* for an objective opinion because M&S can avoid inferences
- *Preparing for change* by answering the "what if" in the design or modification of the system
- *Investing wisely* because a simulated study costs much less than the cost of changing or modifying a system
- *Training more efficiently*, less expensively, and with less disruption than on-the-job training
- *Modifying requirements* for a system design that can be modified to reach the desired goal

It is obvious that there are many uses for and many advantages of using M&S. Not to be overlooked are the disadvantages of M&S. This list is noticeably shorter and includes such things as (1) the *special training* needed in building models, (2) the *difficulty in interpreting results* when the observation may be the result of system interrelationships or randomness, (3) the *elevated cost in money and time*, due to the fact that simulation modeling and analysis can be time consuming and expensive, and (4) *inappropriate use* of modeling and simulation when an analytical solution is best.

CONCLUSIONS

Models are approximations of events, real events as in case studies, or contrived events as in use-case studies. Analysts create models from data; therefore, research for the event or details that go into a case study must be accurate to ensure that the model is sound. With a reliable model, analysts can develop a hypothesis or research question that requires observation of the model. The model is observed via *simulation*, and the simulation can be modified and repeated. Often, models include systems or collections of different elements that together produce results not obtainable by the elements alone. The analyst then conducts an *analysis* of the simulations to draw a conclusion or to verify and validate the research. The ability to apply visualization facilitates the communication or presentation of the model, the simulation, and the conclusions drawn. All of this is *learning by doing*.

The *history of M&S* is quite lengthy, especially if the review starts with the modeled battlefields and wargames of the ancient world. Scientists such as Leonardo DaVinci modeled everything from bridges to life-sized busts to military weapons. DaVinci also tested his inventions and in doing so left us with the first systematic explanation of how machines (or systems) operate. The military engaged M&S to the fullest during the nineteenth and twentieth centuries. By the 1990s, *behavioral modeling* was integrated into military applications of M&S. Behavioral modeling is also present in many other research fields because it focuses on human activity and behavior derived from the psychological and social aspects of humans.

KEY TERMS

model

simulation

visualization

system

behavioral modeling

analysis

experimentation

training

stand-alone simulation

integrated simulation

zero-sum simulation

virtual simulation

statistical modeling

formal modeling

agent-based modeling

REFERENCES

[1] Banks J, ed. *Handbook of Simulation: Principles, Methodology, Advances, Applications, and Practice*. New York: Wiley, 1998.

[2] Fishwick PA. *Simulation Model Design and Execution: Building Digital Worlds*. Upper Saddle River, NJ: Prentice Hall, 1995.

[3] Law AM, Kelton WD. *Simulation, Modeling, and Analysis*, 4th ed. New York: McGraw-Hill, 2006.

[4] IBM Research. Deep Blue. http://www.research.ibm.com/deepblue/watch/html/c.shtml. Accessed Jan. 2, 2008.

[5] Young JP. *History and Bibliography of War Gaming*. Washington, DC: Department of the Army, 1957.

[6] Link simulation and training. http://www.link.com/history.html. Accessed Jan. 2, 2008.

[7] Ören TI. Invited Tutorial: Toward the body of knowledge of modeling and simulation (M&SBOK). In: *Proceedings of the Interservice/Industry Training, Simulation Conference*, Orlando, FL, Nov. 28–Dec. 1, 2005, pp. 1–19.

[8] President Dwight D. Eisenhower's farewell address, 1961. http://www.ourdocuments.gov/. Accessed Jan. 2, 2008.

[9] Spacewar. http://www/wheels.org/spacewar. Accessed Jan. 2, 2008.

[10] Mazewar.   http://www.digibarn.com/history/04-VCF7-MazeWar/index.html. Accessed Jan. 2, 2008.

[11] TACDEW. http://www.ntsc.navy.mil/Programs/TrainerDescriptions/Surface/ TACDEW.cfm. Accessed Jan. 2, 2008.

[12] SIMNET.   http://www.peostri.army.mil/PRODUCTS/PC_BASED_TECH/. Accessed Jan. 2, 2008.

[13] Carley K. Social behavior modeling. In: *Defense Modeling, Simulation, and Analysis: Meeting the Challenge*. Washington, DC: National Academies Press, 2006.

[14] Ören TI. Maturing phase of the modeling and simulation discipline. In: *Proceedings of Asian Simulation Conference 2005* (The Sixth International Conference on System Simulation and Scientific Computing, ICSC'2005), Beijing, China, Oct. 24–27, 2005. Beijing, China: International Academic Publishers–World Publishing Corporation, 2005, pp. 72–85.

[15] Sokolowski JA, Banks CM. *Principles of Modeling and Simulation: A Multidisciplinary Approach*. Hoboken: John Wiley, 2009.

FURTHER READING

Capra, Fritjof. *The Science of Leonardo da Vinci: Inside the Mind of the Great Genius of the Renaissance*. New York: Doubleday Books, 2007.

Clarfield, Gerard H. *Security and Solvency: Dwight D. Eisenhower and the Shaping of the American Military Establishment*. New York: Greenwood Press, 1999.

Langley, P., H. A. Simon, G. L. Bradshaw, and J. M. Zytkow. *Scientific Discovery: Computational Explorations of the Creative Processes*. Cambridge, MA: MIT Press, 1987.

Shenk, David. *The Immortal Game: A History of Chess, or How 32 Carved Pieces on a Board Illuminated Our Understanding of War, Art, Science and the Human Brain*. New York: Doubleday, 2006.

Vernon, Alex, R. Holmes, G. Downey, D. Trybula, and N. Creighton. *The Eyes of Orion: Five Tank Lieutenants in the Persian Gulf War*. Kent, OH: Kent State University Press, 1999.

2 Research Methodologies for Modeling Global Events

INTRODUCTION

As we learned in Chapter 1, *models* are representations of structures or systems, and *simulations* are models with specific inputs.[1] The analyst is most interested in the output of those simulations. We introduced the three common social science modeling techniques in Chapter 1: *statistical modeling*, a traditional method for the discovery and interpretation of patterns in large numbers of events; *formal modeling*, a method that provides a rigorous analytic specification of the choices that actors can make and how those choices interact to produce outcomes; and *agent-based modeling*, a method that allows for the observation of aggregate behaviors that emerge from the interactions of large numbers of autonomous actors. In the social sciences the **inputs** to a model are the attributes needed to make the model match a social environment. The **outputs** to this simulation are the behaviors of the model, and it is these that social scientists seek to analyze, understand, and explain [1].

Because we can seldom work with certainties, we must construct hypotheses that can possess greater or lesser degrees of probability. Simulations allow us to test the validity of these hypotheses and select the best representation. Because we do not live in a uni-directional world

---

[1]Portions of the chapter are based on John A. Sokolowski and Catherine M. Banks, Modeling and simulation: real world examples, in *Principles of Modeling and Simulation: A Multidisciplinary Approach*, by John A. Sokolowski and Catherine M. Banks, eds. New York: John Wiley & Sons, Inc., 2009.

---

where problems are static and can be represented in the linear form

real world → information feedback → decisions

we use models to help us characterize a **circular environment**, in which action is based on current conditions ... which are affected by actions ... which change conditions ... and so on [2]. Figure 2-1 displays this in a very simple format or model.

The model shown in Figure 2-1 has a considerable deficiency in that it does not display the complexity of the decision making nor the learning and influence associated with information feedback—both of which affect decision making. A more sophisticated model is needed to better represent this process. Figure 2-2 depicts this circular environment and goes on to show how a more sophisticated model can contribute to the learning and decision process. Observing the differences between these two models is key to understanding how to model the dynamic structures that are integral to the fundamentals of research topics that are inherent to analyzing global events: human behavior, social networks, and entire societies, all of which are dynamic and organic, thus nonlinear in representation. This is also the first step in understanding the importance of modeling and simulation in the social sciences.

So how can the modeling techniques used by social scientists expand to provide a greater depth of representation of the circular environment in which policymakers, leaders, governments, and others must operate? What

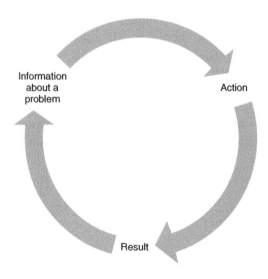

Figure 2-1   Simple decision loop model.

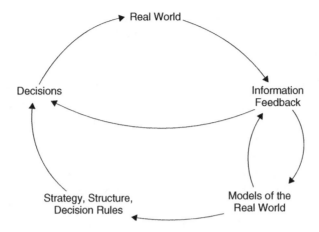

Figure 2-2   Modeling-supported decision loop.

other types of modeling and simulation (M&S) can be applied to topics of study that are integral to the social sciences; topics such as collective security, conflict and cooperation, foreign policy and decision making, international law, international political economy, interdependence, and transnationalism? In this chapter we answer those questions by (1) defining social sciences and its topical areas of research, (2) explaining qualitative and quantitative research and the integration of both types of analyses into models, and (3) discussing the integration of modeling and simulation in social science research.

GLOBAL EVENTS AND THE SOCIAL SCIENCES

Before we discuss global events, let's take a look at the broader category of the social sciences. The **social sciences** are academic disciplines that focus on a study of the human aspects of the world—civilization—emphasizing use of the scientific method in the study of humanity, including quantitative and qualitative methods. There are many disciplines within the social sciences, including anthropology, religious studies, criminal justice, economics, history, political science, sociology, and international studies. The social science disciplines study human society and individual relationships *in* that society as well as individual relationships *to* that society. Understandably, this is a complex network of relationships that must be modeled as a dynamic, organic system. To represent these dynamic, organic systems accurately requires moving beyond quantitative or numeric modeling. To convey *relationships* in a model as more than just a linear link between agents or entities, qualitative analysis must be incorporated because qualitative analysis introduces *ideas, intangibles*, and *subjective data* into the model.

One discipline that focuses on the study of global events is **international studies** (IS). IS is typically an interdisciplinary program of study that incorporates varying degrees of training from a number of disciplines, such as political science, international law, political economy, anthropology, history, psychology, public policy, and management. The discipline strives to prepare students to develop creative responses to current global challenges using historical analogies, encouraging comparative and critical perspectives, and explaining theoretical frameworks for ordering complex data. All of these skills are necessary for analyzing global events. A number of concentrations or themes are examined, such as those at the *Institute of International Studies* at Berkeley, California [3]:

- Peace and Global Security in the 21st Century
- Environment, Demography, and Sustainability
- Globalization, Development, and Human Rights
- Technological Change and the Transformation of the Global Economy

At the *Freeman Spogli Institute for International Studies* at Stanford, students are encouraged to address critically important questions, such as [4]:

- In an increasingly interconnected world, what measures can nations take alone or together to protect against nuclear, biological, or chemical terrorism?
- How can we help current and emerging leaders combat corruption and promote democracy, development, and the rule of law in transitional countries?
- How can health care systems around the world best prevent and treat a range of deadly infectious diseases, including HIV/AIDS, tuberculosis, Avian flu, and potential anthrax or smallpox attacks?
- Will greater economic interdependence among Asian countries reduce the aggressive forms of nationalism currently on the rise in the region?
- How can policymakers generate innovative solutions to the silent killer of our time, global hunger and chronic food insecurity?
- Looking beyond the Kyoto Protocol, how can policymakers develop a credible and effective system to address global warming?

The graduate program in International Studies at *Old Dominion University* (Norfolk, Virginia) focuses on five concentration areas that it deems best suited to prepare students as analysts of global events [5]:

1. *U.S. foreign policy.* This field provides a thorough understanding of all of the concepts and conditions that serve as key components in the making of foreign policy. These factors include, but need not be limited to, economic, cultural, and political considerations, with an emphasis on the fundamentals of foreign policy: the construction, execution, evolution, and implications.

2. *Conflict and cooperation.* This field aims to provide a comprehensive understanding of the various discourses in security studies that critically examine different epistemologies and theories and that apply and test competing explanations of security choices.

3. *International political economy and development.* This field examines the differential production of power and wealth in the world, most notably between the industrialized states of the North and the developing states of the South, by examining the problems of underdevelopment, debt, and dependence. Integral to this concentration is the exploration of alternative strategies for reducing inequalities within and between nations.

4. *Interdependence and transnationalism.* This field explores the main concepts, themes, actors, and processes relevant to the study of transnationalism, interdependence, and power. Transnational problems are cross-national and cross-regional; threats such as terrorism, migration, proliferation, and environmental degradation are examined as well as the impact of transnational forces, including democratization and Islam, the relationship between interdependence and conflict, interdependence and security, global energy, interdependence, and the global environment.

5. *Comparative and regional studies.* This concentration analyzes sociopolitical phenomena cross-culturally and cross-nationally. One dimension of this concentration deals with major sociopolitical issues that are common across national borders and cultures. Such issues include democratization, political violence, regime legitimacy, political change, economic reform, political participation, institutional formation, and so on, all of which can be better understood through comparative study. A second dimension of the concentration is a regional one, which examines closely various regions (or countries) with attention to the aforementioned issues within the regions.

From these examples we can see that IS as a discipline relies heavily on history and political science as foundations to an understanding of critically important questions facing global citizens in the twenty-first century. The discipline also requires an understanding of geography, sociology, and cultural and political aspects to conduct thorough comparative studies. Not to be forgotten are the technical skills necessary to conduct social science research. The wide swath of research topics in the social sciences handily accommodates the incorporation of additional forms of modeling alongside those that are part of general social science methods (statistical modeling, formal modeling, and agent-based modeling). Students utilizing the traditional modeling methods should strive to take those tools to a more advanced level, such as *social network modeling*, which requires characterizing the dynamism and organic nature of a social network. More and more professionals in the social sciences are engaging modeling tools that have their home in the engineering disciplines, such as those discussed in Chapters 3, 4, and 5.

Recall from Chapter 1 the history of modeling and simulation and its uses in various disciplines. In that discussion we also emphasized the importance of qualitative analysis. Studies in geopolitics, political economy, and transnationalism provide some of the variables necessary to affect real-time decision making and policymaking with immediate-, medium-, and long-term goals. It is essential in the development of accurate models depicting these and numerous other factors, all of which comprise a *system* (a construct or collection of different elements required to produce system-level qualities, properties, characteristics, functions, behavior, and performance that together produce results not obtainable by the elements alone). Social science research, combined with advanced modeling and simulation techniques, will best characterize these elements with an eye to representing, understanding, and communicating the complexities of human behavior and real-world consequences. Often, however, there is a disconnect between the research conducted by students in the social sciences and the data necessary for a mathematical formula used in the construction of a model. We will now discuss those differences. Note how the differences range from the assumptions made by the researcher, the purpose and intent of the research itself, the methodology or approach taken by the researcher, and the role of the researcher.

QUALITATIVE AND QUANTITATIVE RESEARCH

In the realm of discovery there are two primary forms of research: qualitative and quantitative. The features of each method are as varied as are the users. For example, Donald T. Campbell, a noted social scientist who spent his career in research methodology, believed that *all research ultimately had a qualitative grounding*.[2] Conversely, Fred N. Kerlinger, another noted social scientist, endorsed a straightforward quantitative approach: "There is no such thing as qualitative data because everything is either 1 or 0."[3] The data reviewed in qualitative research are subjective and open to interpretation by the analyst. Moreover, data such as words, pictures, and behaviors can be assessed differently by two analysts. It is clear that the training and thinking of the analyst affects his or her interpretation of the data. On the other hand, quantitative research involves the analysis of numerical data; therefore, it is objective. The value of a whole number is the same no matter who assesses its value.

Then there are the deductive and inductive approaches to broad investigation: **inductive reasoning**, which allows a theory to be tested via generalized observations, and **deductive reasoning**, which tests the assumption of a hypothesis. Simulation makes possible a hybrid approach to investigation, one that starts with assumptions, then experiments to generate data, and then allows for analysis of the data [1]. It is important to appreciate the major differences between qualitative and quantitative research because students integrating modeling and simulation into social science research need to apply both modes of research.

Qualitative Research

The ancient historian Herodotus approached the study of history by asking questions, looking for answers, and drawing conclusions. This approach to qualitative research abounds in many history texts and narratives based on fact. For example, if one wanted to investigate the everyday lives

[2]Donald Thomas Campbell (1916–1996) was a social scientist (psychology) who concentrated much of his academic career in research methodology. He is best known in the field for coining the term *evolutionary epistemology* and developing a selectionist theory of human creativity. He focused on the study of *false knowledge*, the biases and prejudices that contaminate all sectors of the social network by the perpetuation of erroneous theories by those who have vested interests. Campbell believed that the use of many research approaches, each with its own distinct but measurable flaws, was required to design reliable research projects. His thesis, *Convergent and Discriminant Validation by the Multitrait-Multimethod Matrix*, is frequently cited in social science literature.

[3]Fred N. Kerlinger (1910–1991) earned his Ph.D. in educational psychology from the University of Michigan and spent most of his academic career advancing his ideas on the conduct of research. He believed that research is theory driven while being supported by technical acumen. In 1964 he wrote his best known text, *Foundations of Behavioral Research*. Kerlinger developed and directed a doctoral program in research and measurement at New York University.

of Germans living in the small town of Dachau, Germany, where one of the earliest Nazi concentration camps was built, one would start by identifying sources and information about the city, the operation of the camp, and any people who lived in the city during that time. He would then read the sources and interview people who agree to assist in the study. The analysis comes in the form of editing the written materials (sources and notes) and the interviews to craft the story. The narrative would no doubt be different from that which any individual interviewee shared with the researcher for it would provide a cumulative assessment. It would provide one answer to the question: *What was life like for German citizens of Dachau circa 1940–1944?* Like most researchers, data are accepted at face value unless there is reason to doubt the data. Using the *ask, research, conclude* approach to research requires a researcher to remain ready to modify his or her understanding and interpretation as new materials might put that understanding and interpretation in a new light.

Qualitative researchers conduct information gathering by direct observation, analysis of documents and sources, and interviews. Hence, the need is for smaller focused samples rather than large random samples. Because this type of research facilitates an understanding of behavior, how people feel and why they feel as they do, it requires an interdisciplinary and multidisciplinary field of inquiry. Therefore, qualitative research is primarily exploratory, and by and large it is not conclusive. It is generally considered to be a naturalistic, interpretative approach concerned with understanding the meanings that people attach to actions, decisions, and values within their social network. It also requires understanding the mental mapping process that makes sense of and interprets the social network system.

Qualitative research yields information that is very detailed. Unlike quantitative research, in which volumes of numerical data may also be collected (which can then be *generalized* into an aggregate statistic such as a mean or median), the findings in qualitative research are thought to be "raw data" and are seldom precategorized. The analyst must organize those data. There are a number of ways to do this, as explained in the case studies in Chapters 6 through 9.

The enormous amount of detail in most qualitative research allows for great depth in describing the research topic. Conversely, it is often difficult to determine what the generalizable themes may be. Hence, the most obvious limitations to qualitative research are (1) a plethora of data that are often difficult to generalize, and (2) the subjective interpretation of data rendered by analysts whose training and thinking factor into the conclusion.

Quantitative Research

If the term *quantitative* is defined as the ability to express a finding as a quantity, number, or measurement, quantitative research must represent a systematic investigation of quantitative properties. Quantitative research seeks to develop and engage mathematical models or mathematical theories pertaining to an event. Hence, **measurement** is central to quantitative research. Quantitative research is widely used in scientific and technical disciplines (biology, chemistry, physics, mathematics, psychology, engineering). Simply stated, it is research that examines events through the numerical representation of observations and statistical analysis.

Juxtaposed with qualitative research, which seeks to explain behavior, quantitative research aims to quantify behaviors by measuring variables on which they hinge and intersect, comparing the variables, and pointing out correlations. Those who favor quantitative research over qualitative research argue that quantitative research legitimizes social science research. Those who support qualitative research contend that quantitative research obscures reality—the humanness of an event or social network—by ignoring, underestimating, or neglecting the nonnumeric, nonmeasurable factors. So what is the best research approach?

The authors of this book suggest a combination of quantitative and qualitative data gathering, referred to as **mixed-methods research**. The mix allows for a summarization of large bodies of subjective or qualitative data and a generalization based on objective or quantitative projections. As expertise in the social sciences remains a constant in solving complex world problems, integrating various applications of modeling, simulation, and visualization into social science research is equally important. Modeling and simulation expands and communicates the qualitative analysis of a research topic, and it is grounded in objective, scientific methodology. Engaging modeling and simulation into an empirical analysis allows one to better understand the *what happened* and to explore the *what if*.

Comparisons

There are some generalizations that merit mention regarding qualitative and quantitative research because they speak to the research approach, goals, and uses. As noted earlier, qualitative research analyzes words, pictures, objects, and events so as to provide a complete description of what has been analyzed. Quantitative research focuses on numerical data aimed at counting, grouping, classifying, and then developing statistical models to explain trends, commonalities, and predict or forecast. The quantitative researcher may know what he/she is looking for, therefore can design the study, then collect the data. Conversely, the qualitative researcher typically has an incomplete idea of what he will find or what he is looking for; hence, his design emerges as he progresses through the research. He is not likely to use any tools in his research—he is the **data gatherer**; whereas the quantitative researcher uses many tools, software programs, questionnaires, and so on. Qualitative research can be much more time consuming for the researcher. It will provide a less generalized but very rich and resonant body of research, whereas quantitative research may be more efficient in testing hypotheses and more precise in analysis of targeted ideas but lacks contextual detail. In sum, qualitative research is affected by the process, the unfolding of data, more than with the outcome. *How and why an event occurred* is the research interest; it goes far beyond a simple recording of *what happened*. The reasons that people do as they do, actions associated with human behavior, are the key to qualitative inquiry. Quantitative research can be nondescript. Qualitative research is very descriptive because it is interested in understanding gained through words, pictures, actions, and events. This type of research builds from the details it derives from the research; therefore, the qualitative research process tends to be inductive—building the hypothesis as information is added. The quantitative research tends to be deductive—testing the assumption of the hypothesis. Table 2-1 describes the differences between qualitative and quantitative research modes relative to assumptions, purpose, approach, and researcher role.

MODELING AND SIMULATION OF GLOBAL EVENTS

Today, there are a number of centers that conduct research in the mixed-methods mode. Some are introducing advanced levels of modeling, simulation, and visualization into the research task. At the Massachusetts Institute of Technology (MIT), the *Center for International Studies* (CIS) supports and promotes international research and education. They look for ways to capitalize on MIT's great strengths in science and engineering by examining the international aspects of these fields as they relate to both policy and practice and focusing on issues where science and engineering intersect most closely with foreign affairs [7]. Any researcher striving to understand the highly complex globalized network would be well served to make use of modeling and simulation methods and tools outside the traditional social science modeling toolbox.

As we have seen, qualitative research draws on data that are fluid, often rendering conclusions obsolete; however, models can be adjusted or new

TABLE 2-1   Predispositions of Qualitative and Quantitative Modes of Inquiry

Qualitative Mode	Quantitative Mode
Assumptions	
Reality is socially constructed	Social facts have an objective reality
Primacy of subject matter	Primacy of method
Variables are complex, interwoven, and difficult to measure	Variables can be identified and relationships measured
Insider's point of view	Outsider's point of view
Purpose	
Contextualization	Generalizability
Interpretation	Prediction
Understanding actors' perspectives	Causal explanations
Approach	
Ends with hypotheses and grounded theory	Begins with hypotheses and theories
Emergence and portrayal	Manipulation and control
Researcher as instrument	Uses formal instruments
Naturalistic	Experimentation
Inductive	Deductive
Searches for patterns	Component analysis
Seeks pluralism, complexity	Seeks consensus, the norm
Makes minor use of numerical indices	Reduces data to numerical indices
Descriptive write-up	Abstract language in write-up
Researcher Role	
Personal involvement and partiality	Detachment and impartiality
Empathic understanding	Objective portrayal

Source: Ref. 6. Copyright © 1992 by Pearson Education, reprinted by permission of the publisher.

simulations can adjust for changes in the qualitative data. Qualitative data cannot necessarily be put into a context that can be graphed or displayed as a mathematical term; however, there are ways to translate qualitative data into numerical values. Recall the importance of measurement to quantitative research—measurement can serve to provide the fundamental connection between empirical observation and mathematical expression of quantitative relationships. What are some *tools outside the social sciences toolbox* that can be used to communicate the findings of a well-researched twenty-first century concern? A systems-based approach to modeling is a good place to start.

Systems-Based Approach

In 1993, Kenyon B. DeGreene (Department of Systems Science, University of Southern California) published a textbook entitled *A Systems-Based Approach to Policymaking*, in which he analyzed policy and decision making using a systems-based approach [8]. DeGreene's research suggests engaging a *science and technology of understanding* in observing interactions among people and things (machines and/or events). He bases his understanding on the simple premise that a human being is a complex system—when he or she interacts with another human being (another complex system) or other systems (machines and/or events), the result is an even more complex system. Theoretically, all of these subsystems must perform in a certain manner for the entire system to function. DeGreene has a Ph.D. in physiological psychology; he thinks outside the box by viewing human beings as systems, events as systems, things as systems. This is research and analysis using a **systems-based approach**. It refers to system theories, philosophies, and models and to the concepts and constructs that are the building blocks of those theories, philosophies, and models. This type of research is grounded in quantitative analysis; however, it can be very useful to approach qualitative analysis in a systems-based manner.

Taking this idea of a systems-based methodology a step further is **system dynamics modeling**, introduced by Jay Forrester of MIT's Sloan School of Management. Forrester used computer simulations to analyze social systems and predict the implications of various models. System dynamics deals with the simulation of interactions between objects in dynamic (active) systems. It combines theory, methods, and philosophy to analyze the behavior of systems in a number of domains: management, politics, economic behavior, engineering, and other fields. The key advantage to system dynamics is that it provides a common foundation for understanding and influencing how things change through time [2]. Social science research can utilize system dynamics modeling to create macro-level representations of systems to address the interdependence of the actors, the events, or such variables as history, culture, and religion within a system. In Chapter 3 we provide the theory and methodology behind system dynamics, and in Chapter 6 we apply this modeling paradigm in a case study. Three other modeling methodologies are also explored: agent-based modeling, social networks, and game theory.

A familiar modeling method that we stretch beyond its traditional boundaries is **agent-based modeling**. These are dynamic models that imitate actions and interactions among the *units of analysis* or agents (representing people, organizations, countries—any type of social actor) and the sequence of actions and interactions of the agents over a period of time. Because it is a dynamic mode of modeling (the actors might act in parallel, might be heterogeneous, and might learn from their actions to the model as responsive to the prior action of one or more of the other agents or the environment), the model can become complex given the sequence of behaviors. Added to that is the intrinsically social nature of the model as a result of the actions and characteristics of the agents influenced by the actions and characteristics of the other agents in the social system.

Social network modeling focuses on social behavior; the model takes into account relationships derived from statistical analysis of relational data. This type of modeling facilitates an awareness and understanding of the connections among people, whether they are political leaders, specific groups, and/or cliques in organizations. It also allows for an explanation of a flow of information (useful in determining the effects of propaganda), or the trafficking of small arms, or the identification of outliers or individuals who are isolated from the group (useful in assessing insurgency recruitment efforts). Grouping patterns (algorithms) allow for the separating of large networks into smaller subsets. This draws closer the members, who share identifying marks or attributes. Perhaps they are all from the same religious sect, are of the same age, or have the same political ideology. Integral to social network modeling is analysis or disciplined inquiry into the patterns of relationships that develop and exist among the members in the social system. Also included are the relationships among members at different levels of the analysis (i.e., person to person, group to group, etc.). Because this type of modeling relies on actors who are concrete and observable, the relationships within the social network are usually social or cultural. These types of relationships bind the actors or entities together, making them interdependent entities. In Chapter 4 we elaborate on agent-based modeling and social networks, and in Chapters 7 and 8 we provide case studies examining agent-based modeling and the social network aspects of human behavior modeling.

Game theory modeling is tied closely to the problem of rational decision making. Decision makers, be they politicians, group leaders, military commanders, or chief executive officers, rarely find the ideal or rational solution to contend with every problem they face. Game theory serves as a tool to study the interactions of individuals, also called *players*, in various contexts such as social dilemmas or confrontational situations. The intent of game theory is to observe the interactions between and among the players so that the decision maker can determine what he or she deems to be the best course of action.

There are two types of game theory: *cooperative game theory*, whereby the players can communicate to form winning coalitions, and *noncooperative game theory*, which focuses more on the individual and his handicap of not knowing what the other players will do. Hence, this person has to make his choice among the available options that are not won via coalition building [1]. As you can see, game theory facilitates the ability to analyze strategic behavior where there are conflicts of interest. Social science research can take this traditional modeling method and elevate it to a more sophisticated application. In Chapter 5 we detail game-theoretic modeling and in Chapter 9 provide a case study that expresses an elevated approach to modeling game theory.

MAPPING DATA: A SUGGESTED METHODOLOGY

The following discussion describes a methodology that we developed as a way to model global events [9]. The methodology merges three tools: (1) traditional historical research, (2) a technique for mapping data, and (3) a systems engineering view of complex systems. These three tools combine to provide a unique view of real-world events.

Tool One   Historical research is vital for an understanding of how a particular event has come to fruition. History often reveals the underlying factors that cause and motivate the actions of individuals or groups. Identification of these factors is the first step in understanding the behavior of the system as a whole. These factors also capture the context in which the event developed. Characterizing the context is important because it provides the reasons (cause and motivation) that the event unfolded in the particular manner that it did.

Tool Two   A technique for mapping data is the second tool, and it is the translation into a numerical representation of a qualitative description of the factors capturing the context of an event. Quantitative methods of the direct use of numbers (such as the number of deaths by insurgents or the number of newspaper articles published in favor of a governmental policy) do not provide a contextual significance for these data, which could lead to erroneous causation representations. The method set forth here provides a qualitative-to-quantitative mapping of factors which maintains their context in relation to the event being studied. Each qualitative factor is evaluated independently and scored by one or more subject-matter experts for its relevance and significance to describing the behavior of the system. The scoring may be done using a Likert scale rating, which represents the value of each factor's contribution to the model. As an alternative, a Bayesian assessment may be used to capture this value. Either method maintains the context of how the individual factors influence the event being modeled.

In describing these complex events, scores of factors might have to be considered. Possessing a large number of them can make it difficult for a modeler to map the multiple relationships in a concise manner. Grouping them into categories will facilitate this mapping. Index values representing each grouping can then be developed and used in the modeling process without the loss of contextual relevance.

For example, a study of the insurgency in the Niger Delta resulted in the identification of 187 factors that characterized the insurgency. To make model construction more manageable, these factors were grouped into seven indices as follows:

1. *Polity Index:* a measure of the state's democracy and its effectiveness

2. *Human Rights Index:* a measure of civil liberties abuse and the degree of political marginalization experienced by the civilian population

3. *Social Capacity Index:* a measure of the population's ability to effect social change

4. *Insurgency Index:* a measure of the strength of the insurgency and the factors that affect it

5. *Counterinsurgency Index:* a measure of the will, capability, and commitment of the counterinsurgency effort

6. *Political Economy Index:* a measure of the economy as it applies to rentier states reviewing economic growth, employment, financial conditions, the health sector, inflation, poverty, and trade[4]

7. *External Influences Index:* a measure of the effect that other institutions or states have on one, any, or all of the country's indices

Values for the indices were determined by having a subject-matter expert rate each factor on a Likert scale of −5 to +5, with +5 representing a high positive influence and −5 representing a high negative influence. The ratings were then averaged under the appropriate index to produce a final index value normalized between 0 and 1. These indices were used to seed key parameters in the model, as outlined below. This method preserves the context behind what is actually going on in the country and provides a way to represent this context quantitatively. Table 2-2 illustrates a portion of the factors used for the Human Rights Index and their corresponding Likert scale assessment.

---

[4]*Rentier* is a term in political science and international relations theory used to classify those states that derive all or a substantial portion of their national revenue from the rent of indigenous resources (in this case, oil) to external clients.

TABLE 2-2   Partial Human Rights Index Table

Factor	Measure
The United States provides over $100 million aid to Nigeria, of which $64 million goes to HIV/AIDS	+3
Corruption, rooted in inequality and injustice, is at the intersection of state and society	−2
The human development index is 156/179, and the general development index is 124/179	−3
The Delta region suffers human rights abuses due to government degradation	−3
There is a feeling of collective disadvantage	−3
There is state-sponsored terrorism; the state uses brute force to quash demonstrations	−4

A qualitative and quantitative review of the data also helps link the identified variables with one another in a causal-type relationship. Understanding the influence of one factor on another is an important part of model development. This step is the most difficult to accomplish. One must guard against a correlative connection versus a true causation link. For example, a country's Polity Index is often linked to its ability to resist insurgency. A high Polity Index score indicates that the country has a strong democratic process and has been correlated with a low probability of insurgency. However, the Polity Index looks only at how a government is structured on paper, and not necessarily on how that structure has been implemented on a day-to-day basis. Thus, a country may have a high Polity Index and it may have resisted insurgency, but the Polity Index rating does not necessarily imply that the government structure was a direct or single contributor to stability. One would have to examine the specific nature of the government structure and its effectiveness at satisfying the population in all four areas mentioned above. Once the factors influencing the global event have been identified and their causal links understood, one can begin to develop a conceptual model to represent these factors and links of the method. Here, systems engineering comes into play.

Tool Three   A systems engineering view of complex systems provides various methods for depicting complex systems and displaying the relationships among variables through the use of system dynamics. System dynamics provides a graphical means to represent these variables and their causal relationships through causal loop diagrams [2]. The exercise of creating a system dynamics view of an event necessitates an in-depth translation of a conceptual representation of the event into a more detailed mathematical relationship among the factors.

Choosing the appropriate modeling paradigm will depend on the specific purpose of the model. For example, a system dynamics approach is adequate for exploring a macro-level view of system behavior where sufficient data are available to calibrate the model to existing conditions. Agent-based modeling allows us to represent simple behaviors of actors in a complex system and observe the evolution of that system over time. This type of process may provide insight into the behavior of the system that helps further define a valid representation of the global event. A game-theoretic model may be appropriate for analyzing the event if we are interested in modeling a sequence of decisions made by two or more entities, and the options for these entities are well understood and can be represented by known utility functions. With these three tools in hand, a model can be produced and analyzed. The next step is validation of the findings of the study.

MODEL VALIDATION

Model validation is the process of comparing simulation results derived from a model against the real-world system that the model is meant to represent. In the judgment of the simulation end user, if the simulation results are close enough to the real system, the model is considered a valid representation of the real system or process. "Close enough" is obviously a subjective term that must be interpreted by the person employing the model.

The validation of models of physical phenomena is generally straightforward since the laws that govern those systems are usually well known and mathematically precise. In this case, comparing the simulation results against the real-world system is just a matter of matching them to a 100 percent predictable outcome. Validating models of global events containing social components is more problematic. When modeling global events that have already occurred, one can compare the results of the simulation with the historical account to judge validity. One can then attempt to extrapolate that model to investigate what will happen in the future. This process, known as **predictive modeling**, endeavors to answer a specific question or set of questions. (The reader is referred to Balci's article for a detailed discussion of this type of validation [10].)

The other purpose of developing a model is to explore a system that cannot readily be manipulated in the real-world so as to gain insight into how it might behave or what factors have the greatest influence on that behavior. Modeling and simulation are emerging as key disciplines that will foster this type of experimentation. However, validating this type of model becomes much more difficult since there is no standard result against which to compare it. Here, the modeler and end user must rely on valid inputs to the model and verifiable behaviors within the model as the basis for validity. This will allow for exploration of how the system could possibly behave so that new insights can be learned about the real-world system. In this case, we must guard against treating this type of model as predictive. The modeler must clearly identify under which of these conditions a simulation is being produced to ensure that there is no confusion about prediction versus insight. There are those who will take exception to using modeling and simulation for this purpose. (The appropriateness of this approach has been discussed in detail by Axelrod [11].)

CONCLUSIONS

In this chapter we explored the integration of modeling and simulation in social science research. The circular environment that exists in decision making, in relationships, and even in analysis (as things are discovered, analyses change, and as analyses change, new and/or additional assessments arise) requires modeling in a nonstatic form. Modeling and simulation enable the researcher to characterize the dynamic, organic social network and human behavior that are integral to social science research. The social sciences, particularly the interdisciplinary nature of international studies, lend themselves to the incorporation of additional modeling techniques as well as expansion of traditional modeling methods.

The social sciences are well grounded in qualitative research, which relies on understanding behavior and is exploratory, naturalistic, and detailed. Quantitative research, which relies on measurement, is also important, as many in the social sciences argue that it legitimizes the research. We recommend a mixed-methods research approach to develop models that accurately characterize the subjective (qualitative) context while developing a mathematical foundation of the model (quantitative).

There are well-established modeling tools in the existing social sciences toolbox, some of which can be further developed and expanded to create more sophisticated models. Making use of a systems-based approach to modeling to explore, explain, and analyze human behavior is a recommended place to start. Although systems-based analysis tends to be an engineering modeling methodology, it is endorsed by prominent social scientists, who suggest the existence of a *science and technology of understanding* and an analysis of *behavior as a system*.

This chapter also provided a suggested methodology for mapping qualitative data factors into bins or indices which can then be translated into a numerical representation. The discussion concluded by highlighting the importance of model validation—the process of comparing simulation results derived from the model against the real-world system that the model is meant to represent.

KEY TERMS

inputs	data gatherer
outputs	systems-based approach
circular environment	system dynamics modeling
social sciences	agent-based modeling
international studies	social network modeling
inductive reasoning	game theory modeling
deductive reasoning	model validation
measurement	predictive modeling
mixed-methods research	

REFERENCES

[1] Gilbert N, Troitzsch KL, eds. *Simulation for the Social Scientist*. New York: Open University Press, 2005.

[2] Forrester JW. *System Dynamics and the Lessons of 35 Years*. Report D-4224-4. Apr. 29, 1991.

[3] Institute of International Studies, Berkeley, CA. http://globetrotter.berkeley.edu. Accessed Apr. 1, 2008.

[4] Freeman Spogli Institute Institute for International Studies, Stanford, CA. http://www.fsi.stanford.edu. Accessed Apr. 1, 2008.

[5] Graduate Program in International Studies, Old Dominion University, Norfolk, VA. http://al.odu.edu/gpis/. Accessed Apr. 1, 2008.

[6] Glesne C, Peshkin A. *Becoming Qualitative Researchers: An Introduction*. New York: Longman, 1992.

[7] Massachusetts Institute of Technology, CIS, Cambridge, MA. http://web.mit.edu/cis/wg.html. Accessed Apr. 1, 2008.

[8] DeGreene KB. *A Systems-Based Approach to Policymaking*. Boston: Kluwer Academic, 1993.

[9] Sokolowski JA, Banks CM. From empirical data to mathematical model: using population dynamics to characterize insurgencies. Presented at the Summer Simulation Multiconference, San Diego, CA, 2007.

[10] Balci O. Verification, validation, and testing. In: Banks J, ed. *Handbook of Simulation*. New York: Wiley, 1998, pp. 335–396.

[11] Axelrod R. Simulation in the social sciences. In: Rennard JP, ed. *Handbook of Research on Nature Inspired Computing for Economics and Management*. Hershey, PA: Idea Group, 2007, pp. 90–100.

PART II
Modeling Paradigms

3 System Dynamics

INTRODUCTION

System dynamics is a modeling method used to study complex systems in a methodical manner. Why is an approach like this needed? For all but the simplest systems, it is difficult and sometimes impossible for most humans to comprehend all the variables and their corresponding relationships that make up a complex, dynamic system. What is needed is a method to clearly represent, both graphically and mathematically, the factors that influence how a system behaves and how those factors are related to one another. In this chapter we introduce system dynamics as one of the most effective modeling tools of dynamic, organic systems.

Through the constructs available in system dynamics one can outline the structure of a system. As this outline develops, one may notice relationships among variables that were never considered before. In this manner, system dynamics facilitates learning and gaining insight into the system to define its behavior and operation more clearly and completely.

System dynamics also provides a means to observe the behavior of a system over time through simulation. Simulation may be the only way to study the behavior of most complex systems. Many real-world systems do not lend themselves to experimentation because of the potential consequences. Experimentation may be too costly or too dangerous. Simulation helps eliminate these barriers.

A key concept of dynamic systems is *feedback*, the interaction of one system variable with another. It is the interactions among all the system variables that produce the complex behavior, not the complexity of the system components themselves. Feedback exists in one of two forms, positive or negative. **Positive feedback** tends to reinforce what is happening

Modeling and Simulation for Analyzing Global Events, By John A. Sokolowski and Catherine M. Banks
Copyright © 2009 John Wiley & Sons, Inc.

in a system. An example of this type of feedback was the buildup of nuclear weapons during the Cold War. The more nuclear weapons that were deployed by NATO, the more weapons the Soviet Union developed. This, in turn, led NATO to produce additional weapons.

Negative feedback opposes what is happening in a system and tends to limit or reverse the growth in a system. Consider a country with a limited food supply. As the population of a country grows, more food is consumed. This causes less food to be available, which in turn limits the number of people the food supply can support, which slows (starvation) or stops (death) the population increase.

Even though there are only two types of feedback, there may be hundreds or thousands of feedback loops making up a system. Trying to understand the behavior of a system by focusing on a small subset of these loops would be unwise because this subset may not capture the influence of other parts that constitute the system. This is where simulation becomes important. A computer simulation of a complex system can go far beyond what most human beings can readily grasp, thus taking into account all feedback loops, for a more accurate representation of system behavior. The results of these simulations are often counterintuitive because of our inability to identify interrelationships across the entire system.

Next we discuss the components of system dynamics modeling and the methods used to combine those components into models and simulations that reflect the dynamics of the system under study.

DYNAMIC SYSTEM BEHAVIOR

The behavior of a system follows from its structure, which is dependent on the existence of various feedback loops and the variables that they interconnect. An explanation of the fundamental and derived dynamic behavior is needed to understand the typical behaviors or responses that characterize systems, along with the feedback modes that govern them.

Fundamental and Derived Dynamic Behavior

Systems exhibit six modes of behavior:

1. Exponential growth
2. Goal seeking
3. Oscillation
4. S-shaped growth

5. S-shaped growth with overshoot
6. Overshoot and collapse

The first three modes are classified as **fundamental modes**. The other three derive from the fundamental modes.

Exponential Growth   Exponential growth is a product of positive feedback. The more positive feedback in a loop, the greater the rate of change of a system variable. A common example is money that earns compound interest. The more money that is invested, the more interest is earned, which leads to a greater balance, with still more interest being earned. Figure 3-1 depicts a curve displaying exponential growth characteristics. Pure exponential growth has the property of doubling its output value in a fixed amount of time. That is, it takes the same amount of time to go from $1 to $2 as it does from $1 million to $2 million. This is a direct result of positive feedback. The characteristics of the system determine the value of the doubling time. For some systems it may be very short, for others quite long. Growth can also be in the negative direction. An example of this behavior is a downturn in the stock market, which causes investors to unload their stocks, leading to an even bigger downturn.

Goal Seeking   Goal-seeking behavior is characterized by the state of a system approaching a specific value and leveling off at that value. Figure 3-2 portrays this behavior. Here, negative feedback acts to limit the growth of the system to some desired value. It counters other feedback, which may try, and succeed, in moving it away from this value. As the system responds, its output is compared to the desired state. When a difference exists between that state and the actual value, the system tries to correct itself to achieve that value. As the state of the system approaches

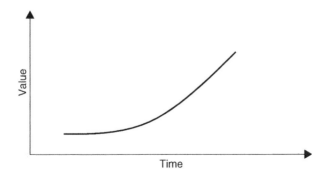

Figure 3-1   Exponential growth curve.

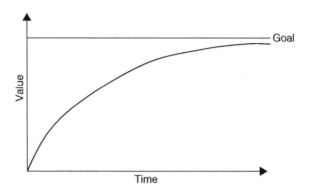

Figure 3-2   Goal-seeking behavior.

the goal, its rate of change decreases so that it does not overshoot the intended value.

Oscillation   Oscillation is similar to goal seeking in that its behavior is governed by negative feedback. The difference is that an oscillatory system overshoots its goal continuously and has to correct itself to once again approach the goal value, all the while overshooting it again. Figure 3-3 shows a typical oscillating system. A factor that contributes to oscillation is **delay**. As some undesirable behavior of a system is observed, changes in one or more parameters are made to try to correct that behavior. It may take some time for the changes to be effective; as a result, the system continues to respond in an undesirable manner. If the value of these delays is not well known, it may be problematic for someone to anticipate when to take corrective action, and that action may come too late. Thus, the system overshoots its desired value.

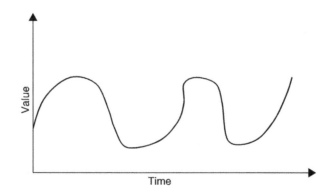

Figure 3-3   Oscillatory system.

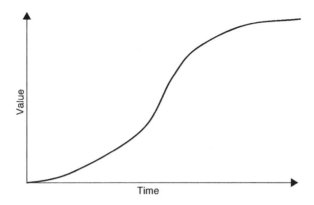

Figure 3-4   S-shaped growth.

S-Shaped Growth   Thus far we have discussed exponential growth, goal seeking, and oscillation. As we pointed out, these behaviors are governed by positive feedback, negative feedback, and negative feedback with delay. Combining these behaviors produces other, more sophisticated behaviors, such as *S-shaped growth*.

It is highly unlikely that a real system will continue to grow or decline forever. There are almost always factors that limit this behavior. Such is the case for S-shaped growth, shown in Figure 3-4. This type of growth starts out as exponential because of positive feedback, but then it levels off at some constant value because of the coupling of negative feedback. One example of such behavior is the *carrying capacity* of an ecological system. The numbers of organisms that inhabit an environment are limited by the resources in the environment and the resource requirements of the organism. Initially, there may be sufficient resources for the organism to grow in an exponential manner. However, as the population increases these resources are consumed at a faster rate, providing negative feedback to the system and causing the growth to level off.

S-shaped growth only occurs when two conditions are present. First, no significant time delays can exist in the negative feedback loop. If they do, the system would overshoot and oscillate. Second, the carrying capacity must be a fixed value. That is, the environment continues to replenish itself to maintain this capacity and it is not itself consumed by the population under study.

S-Shaped Growth with Overshoot   S-shaped growth assumes no significant delay in the negative feedback loop, but if such a delay exists, *S-shaped growth with overshoot* will result. See Figure 3-5 for this behavior.

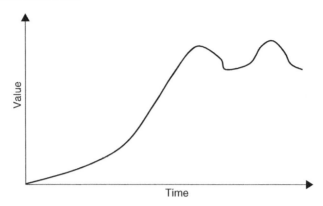

Figure 3-5 S-shaped growth with overshoot.

Overshoot and Collapse *Overshoot and collapse* results when the assumption of a constant carrying capacity is removed from an S-shaped growth system. Here the carrying capacity is either consumed or unsustainable, causing a subsequent drop. This drop forces the system to respond by adjusting its level to what can be supported by the resulting capacity. Figure 3-6 shows such a response. This type of behavior typically occurs when an ecosystem is relatively small and has competition for its resources outside the specific system being studied.

There are also other behavior modes, such as equilibrium, randomness, and chaos, but these modes are not typical of most system responses. Understanding behavior modes is integral for the next step in engaging a system dynamics methodology: the constructs used to build system dynamics models and the combinations of these constructs to produce representations of a vast array of systems.

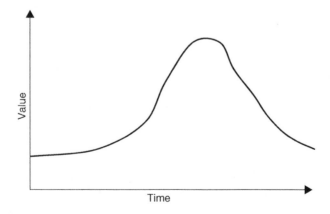

Figure 3-6 Overshoot and collapse.

BUILDING BLOCKS OF SYSTEM DYNAMICS MODELS

This section discusses components and methods used to conceptualize the behavior of a system and then to model and simulate that behavior in a quantifiable manner. It is accomplished by introducing a set of tools used by system modelers that leads through this progression.

Causal Loop Diagrams

Causal loop diagrams are conceptual models of how a system is perceived to behave. As noted above, humans have a difficult time formulating mental models to a high degree of complexity. A causal loop diagram can help formalize the structure of a system through visualization of the relationships of the variables that make up the system. The variables are connected by arrows to show the influences that exist. What begins to emerge is a depiction and recognition of the feedback loops that may not have been apparent in a mental model of the system. Figure 3-7 is a representative causal loop diagram of a simple population system with two feedback loops: a birth rate loop and a death rate loop. In this example, the system variables—birth rate, fractional birth rate, death rate, average lifetime, and population—are connected by causal links. For example, the link connecting birth rate and population is meant to show that birth rate causes some amount of change in the population level. The *plus sign* at the tip of the arrow between these two variables indicates link polarity. This means that if the cause increases, the effect increases above what it should otherwise have been. Similarly, if the cause decreases, the effect decreases below what it should otherwise have been.

A *minus sign* at the arrow tip indicates that if the cause increases, the effect decreases below what it would otherwise have been. It is important to note that link polarities depict how the system is structured; they do not indicate the actual behavior of the variables. They only indicate what would happen if there were a change given that all the other variables remain constant.

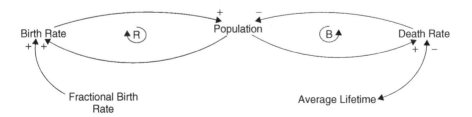

Figure 3-7   Causal loop diagram of a simple population system.

Guide for Developing Causal Loop Diagrams   When developing causal loop diagrams for a system, one must be careful not to confuse causation with correlation. **Correlation** is based on the past behavior of a system. Because two variables changed previously in a specific manner does not mean that one caused the other. The *causal relationship* must be verified to provide an accurate structure for the system. For example, a study published in the journal *Nature* reported that young children who sleep with a light on are more likely to develop myopia in later life. So the natural tendency is to think that sleeping with a light is a direct cause of myopia [1]. A follow-on study by Ohio State University showed no causal relationship between the two. Instead, it pointed out that the cause was hereditary and that parents with myopia tended to leave the light on in their child's bedroom [2]. So there was a correlational relationship but not a causal relationship linked to the light.

Verifying causal relationships can be very difficult. One method employed is scientific experimentation. But one must be able to control the environment to the point of isolating the suspected causal factor. In complex systems this is usually not possible, which is why they are being simulated in the first place. So the modeler must take care to check the causal influences as much as possible when linking variables in a system dynamics model.

One should label loop polarity so that there is a clear indication of how one variable is influencing another in that loop. This usually requires postulating a change in one variable and propagating that change around the loop to see how the affected variable changes. It is also helpful to name each loop in the system so that one can see at a glance its role or contribution to the overall system. It can also facilitate discussion among a group working on the same model by serving as a method of reference.

Including the important delays in the diagram is also required. These provide a marker both for understanding that there is a causal delay between variables and as a precursor to account mathematically for these delays when the full system dynamics model is developed. Figure 3-8 illustrates polarity, delay, and naming methods.

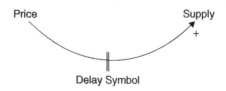

Figure 3-8   Loop polarity, delay, and naming example.

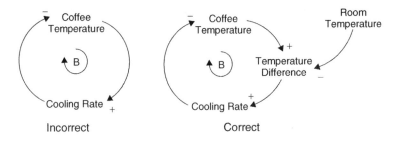

Figure 3-9 Explicit goal examples.

It is important to note that all negative feedback loops have goals, which are the desired state of the system. The modeler should ensure that the goal of each negative loop is explicit so that there is a clear indication or *absolute measure* of what is being compared. Figure 3-9 shows correct and incorrect ways to account for the goal.

Causal Loop Diagram Development Example   The Soviet Union (SU) and the United States were engaged in an arms race from 1945 to 1989. As a result, each side built up a stockpile of nuclear weapons based on the concepts of mutually assured destruction and strategic deterrence. Researchers have used system dynamics to describe how this arms race evolved [3,4]. Let us explore the construction of a representative arms race model by developing a causal loop diagram that describes the variables and relationships that governed this race. The diagram is developed following a three-step process:

1. Problem definition
2. Key variable identification
3. Causal diagram development

Problem Definition   The task is to develop a causal loop diagram that accounts for the behavior of the Cold War arms race between the SU and the US following the guidelines noted above. The initial strategy followed by the two countries was to stockpile nuclear weapons to a level such that neither country had a clear advantage over the other and that if one country conducted a first strike against the other, the country attacked could respond and deliver an equivalent amount of destruction. The result was a rapid (exponential) buildup of nuclear weapons in both countries. As the level of weapons increased, both countries realized that they could not continue this buildup unchecked. Talks and subsequent treaties were therefore negotiated to limit the number of nuclear warheads that each country possessed. This description will serve as the basis for our model.

Key Variable Identification   From the problem definition above, several variables are identified.

1. *U.S. production rate:* the rate at which the United States can produce nuclear weapons (weapons/year)
2. *Soviet production rate:* the rate at which the Soviet Union can produce nuclear weapons (weapons/year)
3. *U.S. arms:* the total number of nuclear weapons possessed by the United States (weapons)
4. *Soviet arms:* the total number of nuclear weapons possessed by the Soviet Union (weapons)
5. *Threat to Americans:* a measure of how much the United States feels threatened by the number of Soviet arms (scale 0 to 1)
6. *Threat to Soviets:* a measure of how much the Soviet Union feels threatened by the number of U.S. arms (scale 0 to 1)
7. *U.S. SALT limit:* the number of warheads the United States is allowed to have according to the Strategic Arms Limitation Talks (SALT)
8. *Soviet SALT limit:* the number of warheads the Soviet Union is allowed to have according to SALT

Causal Diagram Development   With the key variables identified, the causal loop diagram can begin to be formulated. Note from the problem statement that the initial policy followed by both countries was to build weapons to meet the perceived threat. This behavior is indicative of a reinforcing loop, which produces exponential growth. The causal loop diagram of Figure 3-10 captures this behavior. Left unchecked, the number of U.S. and Soviet arms would continue to grow exponentially as shown in Figure 3-11.

There is one balancing factor that limits this exponential growth. The problem statement indicated that treaties were negotiated to place limits on the number of warheads that each country could possess. So the treaty limit enforces a goal-seeking negative feedback loop that causes the growth in the number of weapons to slow as it approaches this limit.

What is the SALT limit? For purposes of this example, each country will tend to build weapons until its treaty limit is reached. As the number of weapons reaches that limit the production rate decreases until the limit is exactly matched. Figure 3-12 depicts how this behavior affects the

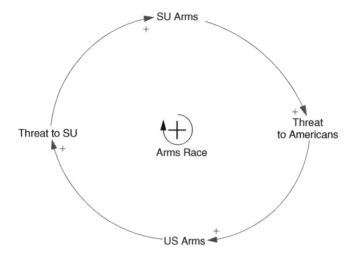

Figure 3-10   Basic arms race causal loop diagram.

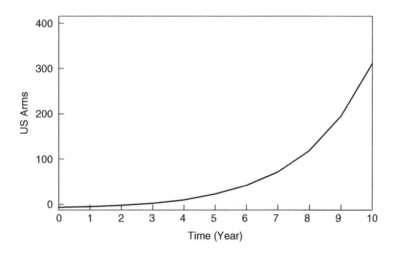

Figure 3-11   Exponential arms growth.

number of US arms. A similar loop can be constructed for SU arms. Note that this loop is a negative feedback loop and tends to limit the number of US arms as the *limit gap* approaches zero.

With all variables accounted for, the next step is to assemble them into the finished causal loop diagram for the entire system. That diagram is shown in Figure 3-13. Two feedback loops merit attention. The first is

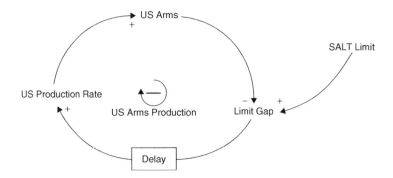

Figure 3-12   US arms production balancing loop.

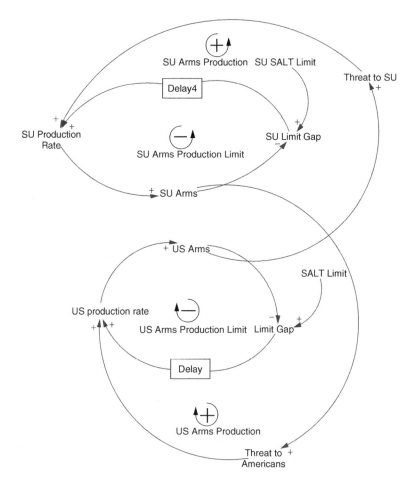

Figure 3-13   Complete arms race causal loop diagram.

the *arms production loop*. This loop is similar to the basic driving loop depicted in Figure 3-10, which was responsible for the exponential growth in the number of weapons for each country. It is tempered by the SALT limit imposed on each country. The resulting response to the overall system is a goal-seeking growth curve that drives the number of weapons to the treaty limits.

Causal Loop Diagram Limitations   Causal loop diagrams are important for depicting and making sense out of complex systems. They are a tool for visually representing the many interconnections that exist in such systems. Keep in mind, however, that these diagrams are simplifications of the actual system. By its very nature, M&S does not capture every aspect of a real-world system. These diagrams are also not final. As the modeler learns more about a system from a modeling and simulation effort, the diagram should be modified to account for newfound knowledge. Causal loop diagrams also do not specifically capture the quantitative aspects of a system because they do not accumulate levels for the variables under consideration, and no mathematical formulas link the variable responses. These two issues are addressed in the next section with the introduction of **stock-and-flow diagrams**, which make it possible to convert a causal loop model into a working simulation of the system.

Stock-and-Flow Diagrams

As pointed out above, causal loop diagrams do not allow for the response of a system over time and the accumulation of variable totals. Stocks and flows provide such a means. **Stocks** are the accumulators of the systems; **flows** represent the rate of change of variables going into and coming from the stocks. Stocks represent the state of the system. At any given point in time, the stock levels describe exactly what exists in the system.

From a diagramming standpoint, stock-and-flow diagrams have four components, shown in Figure 3-14. Stocks are represented by rectangles that possibly have one or more inflow and outflow components. The large arrows represent these flows and show their direction. Valves on the arrows act much like valves in a fluid system. They regulate the flow of input and output variables and represent the rate-of-change factors for the stocks. The **cloud pictures** are sources and sinks, and they account for stock variables that are outside the system being studied. They either feed the system or receive output from the system under study.

From a mathematical perspective, stocks and flows are represented by differential equations that describe the rate of change of each

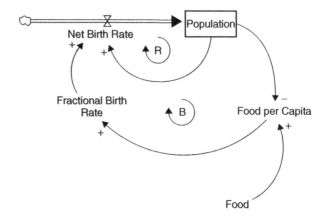

Figure 3-14   Stock-and-flow diagram.

stock in terms of its flows. A general differential equation for this representation is

$$\frac{dS}{dt} = \text{inflow}(t) - \text{outflow}(t) \tag{3-1}$$

Here S represents the stock level and dS/dt represents the rate of change of S. Integrating equation (3-1) allows for calculation of the stock level at any given time:

$$S(t) = S(t_0) + \int [\text{inflow}(t) - \text{outflow}(t)]\, dt \tag{3-2}$$

Stocks are always the source of delays in a system. The delay manifests itself in terms of the difference between the input and output of the stock. This difference can create a lag in the system, which in turn influences how the other system components respond.

From a consistency standpoint it is important to identify the units of measure for stocks and flows. This helps ensure that the interconnecting variables are properly linked and that they support one another mathematically.

Stock-and-Flow Diagram Example   Recall the arms race example from Figure 3-13. This diagram will be used to derive the respective stock-and-flow model. The first step in this derivation is to identify the stocks for the system. In this case the system is concerned with the number of weapons in each country's arsenal, so this becomes the stock

Figure 3-15   Initial arms race stock-and-flow diagram for U.S. arms.

that the model will accumulate. Each country's accumulation of weapons is controlled by the arms production rate. Figure 3-15 shows the initial stock-and-flow diagram for the U.S. portion of this system.

There are no other stocks for this model since the only variable of interest from the standpoint of the total amount is the number of arms for each political entity. The remaining variables are used to calculate each of the production rates. Figure 3-16 provides a complete stock-and-flow diagram for this system.

Now that all elements of the stock-and-flow diagram are identified, mathematical relationships among the variables can be defined. These relationships will provide the basis for the temporal evolution of the model in the form of a simulation. The mathematical portion of the U.S. arms diagram will be illustrated. The Soviet portion is mathematically identical. We start with the equation for the U.S. arms stock, identified by the variable USARMS, and the U.S. arms production rate, USARMSPR.

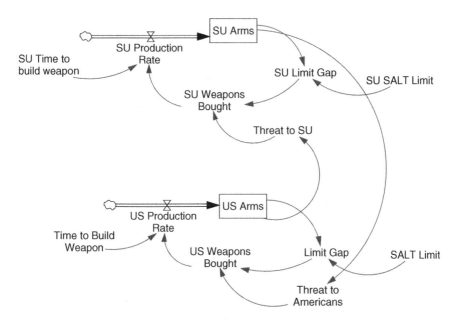

Figure 3-16   Complete stock-and-flow diagram for the arms race model.

$$\text{USARMS}(t) = \text{USARMS}(t_0) + \int (\text{USARMSPR})\, dt \qquad (3\text{-}3)$$

$$\text{USARMSPR} = \frac{\text{USWeaponsBought}}{\text{TimeToBuildWeapon}} \qquad (3\text{-}4)$$

$$\text{USWeaponsBought} = \text{int}(\text{LimitGap} \times \text{ThreatToAmericans}) \qquad (3\text{-}5)$$

$$\text{ThreatToAmericans} = 1 - \frac{1}{\text{USSRArms}} \qquad (3\text{-}6)$$

$$\text{LimitGap} = \begin{cases} \text{SALTLimit} & \text{SALTLimit} - \text{USArms} > 0 \\ 0 & \text{SALTLimit} - \text{USArms} \le 0 \end{cases} \qquad (3\text{-}7)$$

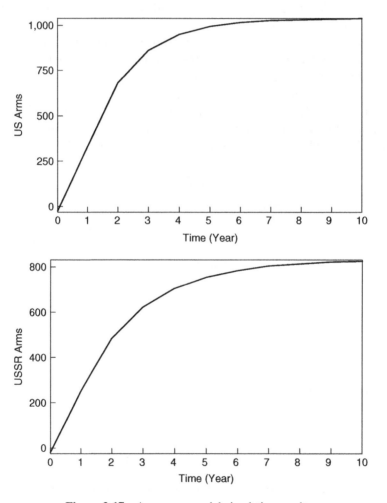

Figure 3-17   Arms race model simulation results.

There are other possible mathematical representations of these variables. However, the ones described by equations (3-3) through (3-7) provide a reasonable set with which to begin exploring the dynamics of the arms race model.

A fictitious set of values was chosen to run a simulation of this model. The results of this run are shown in Figure 3-17. The overall response of the system is consistent with the previous discussion on the expected behavior of this system, with its given feedback loop structure. The reader is encouraged to implement the model for himself and explore possible policy alternatives and their effects on the system.

CONCLUSIONS

This chapter described system dynamics as a means to capture the dynamic behavior of complex systems by looking at the variables and the relationships that contribute to this behavior. Fundamental and derived behaviors were presented along with how those behaviors were generated through various modes of feedback. The concept of causal loop diagrams was introduced as a method to portray graphically system relationships and the feedback loops that control overall system response. Finally, a method to derive stock-and-flow diagrams from causal loop diagrams was presented along with the mathematical representation of stocks and flows that allows for system simulation.

KEY TERMS

system dynamics   correlation
positive feedback   stock-and-flow diagram
negative feedback   stocks
fundamental mode   flows
delay   cloud pictures
causal loop diagram

REFERENCES

[1] Quinn GE, Shin CH, Maguire MG, Stone RA. Myopia and ambient lighting at night. *Nature* 1999;399(6732):113–114.
[2] Zadnick K, Jones LA, Irvin BC, Kleinstein RN, Manny RE, Shin JA, Mutti DO. Vision: myopia and ambient night-time lighting. *Nature* 2000;404(6774): 143–144.

[3] Ward MD. Modeling the USA–USSR arms race. *Simulation* 1984;43: 196–203.

[4] Ruloff D. The dynamics of conflict and cooperation between nations: a computer simulation and some results. *Journal of Peace Research* 1975;12(2):109–121.

4 Agent-Based Modeling and Social Networks

INTRODUCTION

An agent-based model is an important tool for investigating many types of human and social phenomena associated with global events. The concept of an agent-based model has its origin with John von Neumann and his von Neumann machine, a theoretical machine that could follow directions to produce a copy of itself.[1] The important idea here is that of a computer being able to create a complex system on its own by following a set of rules or directions and not having the complex system defined beforehand by a human being. Several other scientists followed in von Neumann's footsteps, including John Conway and Craig Reynolds. Conway constructed the Game of Life, which uses agents to represent people in a neighborhood to see how the neighborhood evolves over time given a set of four simple rules. Reynolds created Boids, which represented a biological system containing birds to explore their flocking patterns.

In the context of this chapter, **social networks** are conceptual models of social systems that help represent the structure of a system and the relationships of the entities that make up the system. They are often depicted in graph form and they are a convenient way of representing the connections among agents in an agent-based model. Let's explore the specifics of these two concepts.

---

[1]John von Neumann was a Hungarian-American mathematician who is credited with major contributions to many fields, including economics, game theory, computer science, and statistics.

Modeling and Simulation for Analyzing Global Events, By John A. Sokolowski and Catherine M. Banks
Copyright © 2009 John Wiley & Sons, Inc.

AGENT-BASED MODELS: DESCRIPTION AND DEFINITION

To begin the discussion of agent-based models, a definition is in order. First, these models consist of **agents** that are defined as *autonomous software entities that interact with their environment or other agents to achieve some goal or accomplish some task*. This definition has several important elements to recognize. Probably the most important of these elements is the concept of **autonomy**. This characteristic is what sets agents apart from other object-oriented constructs in computer science. Agents act in their own self-interest independent of the control of other agents in the system. That is not to say that they are not influenced by those other agents or by their environment; significantly, they do not take direction from other agents. Because of this autonomy, each agent decides for itself what it will do, when it will do it, and how it will be done. These decisions are based on behaviors incorporated into the agent by its designer.

An agent's **environment** and the existence of other agents in that environment also play a key role in how an agent may behave. An agent is embodied with the ability to sense its environment, which includes everything it is aware of external to itself except other agents. As it senses this environment, it may respond to changes in it or may just observe the changes waiting for a specific event to take place. It is also aware of the other agents. It may monitor what they are doing and may communicate with them to request that they accomplish some task; or, it may respond to a request that it has received. This closely represents how a human being interacts with his surroundings and the other persons in it.

Finally, an agent acts to achieve some **goal** or accomplish some task. A task may be to retrieve a piece of data from a specific source or move to a certain location in virtual space. Task accomplishment is generally reactive in nature and does not require a complex set of reasoning to carry out. Figure 4-1 depicts a reactive agent environment. To achieve a goal,

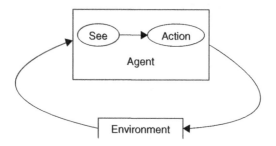

Figure 4-1 Reactive agent model.

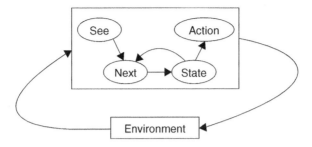

Figure 4-2   State-based agent model.

on the other hand, generally requires more sophisticated reasoning. An example goal may be to accumulate a certain number of objects from the environment. To accomplish this, the agent may have to perform a series of tasks, either in coordination with or in direct conflict with other agents and its environment. Goal achievement usually requires more than just a reactive response to stimuli. It often requires the agent to remember past events in its environment, establish a plan to achieve the goal or goals, and carry out a sequence of tasks in a specific order that leads it closer to the goal. Agents achieve memory by maintaining a representation of the *state* of the world in a **state-space model**. Figure 4-2 represents an agent that maintains awareness of its environment. Here, *state* refers to *a collection of variables that completely describe the virtual environment in which the agent resides*.

An agent may have more than one goal. In this case it must prioritize them, decide what resources it wants to commit to each goal, and devise a plan to achieve as many of the goals as possible. At times, goals may conflict with one another. An agent must then decide which goal to pursue and what actions will lead it closer to all goals. This goal adjudication process is much like how human beings resolve conflicts in deciding what goals they want to achieve through a compromise mechanism.

Agent Reasoning

Because agents are autonomous, they must have some mechanism to reason relative to their environment to decide on how to act. This **reasoning process** forms the heart of agent behavior. Several computational methods have been devised to provide agent reasoning. The more common of these methods are described below.

Finite-State Machines and Markov Chains   **Finite-state machines** (FSMs) are composed of a series of states linked together by transition

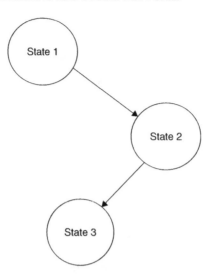

Figure 4-3   Finite state machine.

functions. The states represent all possible situations in which an agent may exist. The transition functions specify under what conditions an agent moves from one state to another. These transitions take place when specific events occur in the agent's environment. Figure 4-3 represents an FSM. The circles depict the states, and the lines connecting them, the transitions.

FSMs work as follows. An agent begins in an initial state. As events occur in its environment, the agent moves from one state to another. As it arrives in a new state it may carry out a specified set of actions defined by that state. These actions may trigger other events in the environment that may affect other agents and may cause them to change state. In this manner, complex behaviors may evolve as agents move from one state to another and trigger a cascading sets of events. Events can also be time-based. An event may be generated after four units of simulation time passes. This could represent an agent having to wait a specified amount of time before it can carry out further actions.

A simple example of an FSM-controlled agent could be that of a traffic light. Suppose that an agent is developed to control the functioning of a standard traffic light such as those found in thousands of cities worldwide. The agent can find itself in one of three states: red, yellow, or green. It can transition from one state to another based on a fixed amount of time passing (time event) or the detection of a vehicle arriving at its intersection (arrival event). Figure 4-4 shows a traffic light FSM.

A **Markov chain** is a variation of an FSM where the transition from one state to another is a probabilistic function. For example, suppose that

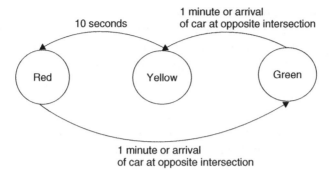

Figure 4-4   Traffic light finite-state machine.

an agent has three states and begins in state 1. When a triggering event occurs, the agent transitions from this state to state 2 with a probability of 1.0. Once in state 2 the agent has an 80 percent chance of returning to state 1 and a 20 percent chance of moving to state 3. One determines the state transition probabilities from known or estimated behavior of the system. Probabilistic transitions allows for the probability to control how the agent behaves based on known probability distributions that characterize these transitions. Figure 4-5 is an example of a Markov chain.

Two other points about FSMs are in order. First, they are memoryless systems in that the actions taken in an existing state do not depend on what state(s) the agent was in before it transitioned to the current state.

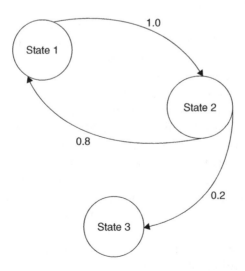

Figure 4-5   Markov chain example.

Therefore, an agent does not need to maintain a sense of history to implement an FSM. Second, an agent developer must be able to identify all possible agent states that govern the desired behavior. If this is not possible, an FSM is not suitable for implementing the agent's behavior. Also, as states are added to represent the desired behavior, the complexity of the agent can grow exponentially. The agent's behavior may quickly become intractable to implement in this manner if the state space is too large. Given this limitation, FSMs constitute a reliable computational technique for implementing agent behaviors that are relatively simplistic.

Rule-Based Models (Expert Systems)   **Rule-based models**, also known as **expert systems**, are very much what their name implies, a set of rules that govern the behavior of an agent. These rules are usually in the form of a set of *if−then−else clauses* that specify *if* this condition or these conditions exist, *then* do the following, or *else* do these actions. They are often referred to as expert systems because they attempt to mimic the set of rules that experts intuitively process when carrying out some behavior or making some decision. One of the very first implementations of an expert system was MYCIN, a medical diagnosis aid. MYCIN assists doctors unfamiliar with microbial infections to prescribe drugs for blood infections [1]. Through a series of questions it asks the doctor about specific symptoms present in the patient. Based on the doctor's response it triggers rules to narrow down the type of infection and prescribe the proper drug. The rule sets were developed through interviews with blood infection specialists who were able describe the reasoning process in sufficient detail to allow for a robust set of rules to be developed for accurate diagnoses under varying inputs and conditions. This process of eliciting expert information is known as **knowledge engineering**.

Typically, there are many rules that an agent must process before arriving at an action or decision. Each rule tests for a specific condition, and when rules are found to satisfy all necessary conditions, a specific action is then found. Two methods exist to process a sequence of rules, forward chaining and backward chaining. **Forward chaining** is referred to as a data-driven approach because it starts with a known set of data about the situation and proceeds toward an action. An inference engine that implements forward chaining searches its rule base until it finds an *if clause* that is true. It then concludes that the *then clause* is true and adds it to its data. Next, it takes the data from the then clause and uses it to find the following "if" clause that is true based on these new data. It continues this process until the desired action or goal is reached.

Backward chaining, also known as *goal-driven chaining*, starts with a list of goals or desired actions and attempts to conclude if one or more of them are true. It searches the rule set for "then" clauses that match the desired goals. If the associated "if" clause is not known to be true, it adds that "if" clause to its data set and searches for a "then" clause that matches the new piece of data. It continues this process until it finds an "if" clause that is known to be true, thus achieving its goal or verifying that it has the appropriate action to carry out. Here is an example of backward chaining using the traffic light system from above. Suppose that the traffic light rule base contains the following rules:

1. If the light is red and 1 minute has passed in this state, turn the light green.
2. If 1 minute has not passed since the last state change, wait for 1 minute to pass.

Suppose that the goal is to get to a green light. The agent would search its rule base, and rule 1's "then" clause would be found to match the goal. The agent checks the "if" clause and finds that "light is red" is true but "1 minute passed" is false. It adds the 1 minute to its data set and searches for a "then" clause that matches wait for 1 minute. It finds rule 2 and checks to see if 1 minute has passed. If not, it continues to wait for this rule to fire. Once the "if" clause of this rule becomes true, its goal has been reached and it can turn the light green.

As one can surmise, it can take thousands of rules to define the behavior logic of an agent. This process could again become intractable when trying to build a very complex behavior structure. A problem also exists if an agent experiences a situation for which there is no rule set that can be executed. Then the agent logic may fail or a default rule may have to be executed which says that the agent cannot understand the situation. The modeler should be aware of these limitations as he designs agent logic using this computational method.

Case-Based Reasoning **Case-based reasoning** uses a set of known cases as a form of memory or experience for an agent. These cases can be thought of as a storehouse of previous solutions to specific situations. They are used as a starting point to reason about new problems that have similar characteristics. For example, an auto mechanic who repairs an automobile by recalling the method he used on a similar repair is employing case-based reasoning. A case contains three main parts: (1) problem or situation description, (2) solution, and (3) outcome. Outcome is important because the solution may not have led to exactly the result desired. But by retaining this case one can compare the fitness of cases to the current problem.

This form of agent behavior definition has been captured in a set of four steps:

1. *Retrieve.* Given a new problem, find cases in the set of past cases that are closely related. Cases are matched by creating an index value that characterizes the new case and comparing that index value to those associated with the cases stored. It is the ability to characterize each case accurately by an index value that is important in this retrieval process.
2. *Reuse.* Map the previous solutions to the new problem to define a set of possible actions or decisions.
3. *Revise.* Choose the closest match from the set of possible actions or decisions and test it on the new problem to see if it is a satisfactory solution. Revise the match if necessary to obtain a solution that better addresses the new situation.
4. *Retain.* Save the new case in the agent's memory for future use.

This method of defining agent behavior has an advantage over FSM and rule-based models in that it allows for finding a possible action or decision even when given incomplete information. Recall that this was not possible for the previous two reasoning methods.

Artificial Neural Networks   An **artificial neural network** (ANN) is a computational framework that mimics the way a human brain stores and retrieves information. An ANN is composed of nodes called *neurons* that operate on a set of inputs mathematically to provide an output. The neurons are arranged in layers, with each layer linked to the layer above and below it much the same as human neurons are linked together in the brain. Figure 4-6 represents a typical artificial neural network.

Figure 4-6 consists of three layers: an input layer, a hidden layer, and an output layer. The middle layer is referred to as a hidden layer because it is solely contained between other layers in the network. Each layer is connected to the next by a set of weighted connections $w_{i,j}$ that are capable of storing information based on the individual weights assigned to each connection and on the mathematical function that each neuron uses to process its weighted inputs. Inputs to an ANN can represent many types of information, such as the shading of each pixel in a digital picture or the characteristics of an electronic signal from a heart monitor. Its output may represent the identity of the person in the digital picture or a warning of an abnormal heartbeat. ANNs are particularly useful in pattern recognition, as pointed out by the two examples above.

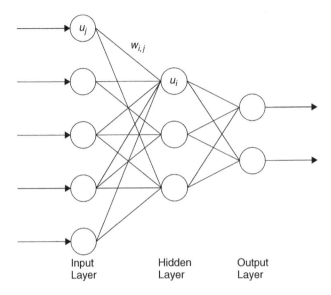

Figure 4-6   Artificial neural network.

Each neuron sums the value of its weighted inputs. If the sum of those inputs reaches a particular threshold, the neuron fires and produces an output signal. This output signal is then fed to other neurons, causing them either to activate (fire) or to produce a final output signal. It is the pattern of these firing neurons for any given input that produces a unique output for the ANN. ANNs with multiple hidden layers containing any number of neurons may be constructed. These networks are capable of storing very large sets of data describing arbitrarily complex relationships among many sets of data.

Once an ANN is constructed it must be trained to create specific behaviors or decisions for a given set of inputs. This training is typically accomplished through training data consisting of inputs and their associated outputs. Inputs are supplied to the network, resulting in an output. That output is compared to the output expected and an error is calculated based on their difference. This error is propagated back through the network via a gradient descent algorithm and the connection weights are adjusted to minimize the error. This process is repeated until the error is reduced to an acceptably small value. This network training method is known as **back propagation**. Once the network is trained with sufficient data to cover the plausible set of expected inputs, it should provide proper output when input data are presented to it. A significant advantage to an ANN is its ability to take incomplete or distorted input data and still produce an output that is similar to one that would have resulted from perfect input data. One disadvantage of an ANN is its inability to determine how a specific output is generated for a certain input. This process is buried in the weights and the individual node activations and cannot be discerned readily by a human observer, and thus an agent's action or decision cannot be explained in a straightforward manner.

Fuzzy Logic and Fuzzy Inference Systems To understand fuzzy logic and fuzzy inference systems, one must start with the foundation on which they are built. That foundation is the **fuzzy set**, a modification of classical set theory that allows for a degree of membership of elements in a set. In classical set theory an element either belongs to the set or does not belong. These sets are also known as *crisp sets*. With a fuzzy set, a member can belong based on a partial relationship. Here is an example to illustrate. Suppose that we have three sets that describe women's height. A woman belongs to the tall set if she is 5 feet 7 inches or greater in height. If she is between 5 feet 4 inches and 5 feet 7 inches tall inclusive, she is in the medium height set. A woman less than 5 feet 4 inches tall is in the short set. Figure 4-7 is a graph representing these three sets. Notice there is no overlap on heights between the three sets. Each person is 100 percent in one of the height categories and 0 percent in the others.

But human beings rarely categorize attributes in such absolute terms. Each person has his or her own perception of what is tall or short, and that perception is not often given in absolute height measurements. A person may observe another and remark that "she is sort of tall," meaning that she is probably more tall than short. So, in this case, there is some

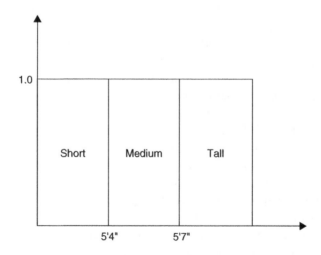

Figure 4-7 The set of women's heights.

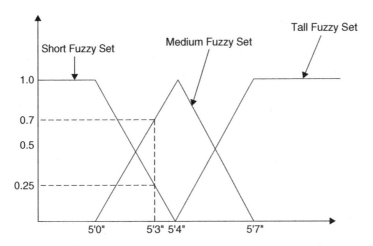

Figure 4-8   Fuzzy set representation of height.

degree of membership in the tall set, but it is not absolute membership. Figure 4-8 is a fuzzy set representation of the women's height categories. Note that there is overlap between the sets, indicating that a person could belong to one or more fuzzy sets with some degree of certainty between 0 and 1. For example, a woman who is 5 feet 3 inches tall could belong to either the short or the medium set, with a degree of belonging 0.25 or 0.7, respectively.

The curves defining these fuzzy sets are known as *membership functions* because they can be used to compute the degree of membership in a set. Membership functions can take many shapes. Some of the more commonly used shapes are triangles and trapezoids. But continuous functions may also describe these sets. Fuzzy sets help relate how humans perceive their world and are important tools for building agent behaviors that mimic that of their human counterparts.

In classical set theory *and, or*, and *not* logical operators allow one to combine elements of sets together. If *A* and *B* are two sets, *A and B* represents all the elements of *A* and *B* together. Fuzzy sets also have these types of logical operators, which allow for the combination of fuzzy sets. Fuzzy sets and their corresponding fuzzy operators are the elements that make up **fuzzy logic**. Just like the rule-based systems described above, fuzzy logic employs *if–then* statements but uses fuzzy sets and fuzzy operators for *antecedent* and *consequent* in these statements. The following is an example of a fuzzy logic statement:

If the sky is overcast and the barometer is falling, then rain is likely.

Overcast and *falling* are linked together by the *and* operator and are fuzzy sets in the antecedent of the statement. *Rain is likely* is the consequent and can be described by a combination of the two fuzzy sets in the antecedent. The process of formulating a mapping of fuzzy inputs to fuzzy outputs using fuzzy logic is known as **fuzzy inference**. A set of these fuzzy *if –then* statements makes up a fuzzy inference system [2].[2] One can see that fuzzy logic can be used to construct an agent reasoning system that closely resembles how human beings think about information. It is for this reason that fuzzy inference systems are a valuable tool for agent-based models.

Agent Communication

Recall that agents in agent-based models are autonomous entities. However, they may interact with other agents in their environment to accomplish their assigned goals. This interaction necessitates being able to communicate with one another to exchange information, coordinate tasks, ask for something, or respond to another agent's request. For this reason some type of agent communication protocol is necessary to allow for these types of interactions. Any user-defined protocol may be used for agent communication. These protocols can be defined specifically for the set of information that must be passed for a given agent implementation. However, several protocols have already been developed and can be adapted to most requirements. Two of these protocols are described here.

The first agent communication protocol to be presented is known as Knowledge Query Manipulation Language (KQML). KQML was developed as part of the Defense Advanced Research Projects Agency's (DARPA) *knowledge-sharing effort*, undertaken in the early 1990s. KQML defines a format for agent messages. This format contains two parts: a *performative*, which may be thought of as the *type of message being sent*, and *parameters*, which contain the *content of the message*. Each parameter is an attribute and value pair. KQML contains a set of forty performatives from which an agent may choose. Each performative is a specific request for a receiving agent to consider. Based on this request and the values of the associated parameters, the receiving agent must decide how to respond. It can choose to act on the message or even ignore it if it does not help it achieve its desired goals. In this manner,

---

[2]The mathematics of computing the value of the consequent based on its antecedents is beyond the scope of this chapter. The reader is referred to Dubois and Prade [2] for a discussion of this process.

each agent maintains its autonomy by choosing on what to act rather than having another agent direct it to carry out some function or task [3].[3]

The second *agent communication language* (ACL) presented here is known as the FIPA ACL. FIPA stands for the Foundation for Intelligent Physical Agents, which developed this ACL in 1999 [4]. This ACL is similar in structure to KQML, with a set of performatives (21) and attribute and value parameter pairs. FIPA improved on KQML by providing comprehensive formal semantics to their language based on formal speech acts theory. Either of these two languages may be adapted to cover most agent communication requirements.

Agent Negotiation

Because agents are autonomous, and because they often have differing goals, they must have a mechanism with which to negotiate and cooperate. As Wooldridge points out, agents in a multiagent system may have been designed and implemented by different people, with different goals. Thus, these agents may not share common goals, which means that encounters between agents more closely resemble games, where agents must act strategically to achieve the outcome they most prefer. Wooldridge also emphasizes the fact that agents are assumed to be acting autonomously therefore they need to be able to coordinate their activities dynamically and cooperate with others [3].

It is for these reasons that agents need negotiation methods to resolve conflicting actions and thus maximize their goal achievement possibilities. There is an extensive body of literature, known generally as *cooperative distributed problem solving*, that describes methods that can be employed in designing negotiation and cooperation schemes, including such things as:

- Task sharing and result sharing
- Handling inconsistency
- Coordination
- Planning and synchronization

These agent encounters could be thought of as games, and therefore formal methods of *game theory* could be used to implement the negotiation and cooperation functionality. Agent-based models often contain multiple agents that form a network of relationships called social networks.

---

[3]The reader is referred to Wooldridge [3] for examples of KQML messages.

SOCIAL NETWORKS

By definition agents are social in nature since they must interact with each other in some manner. To facilitate this interaction the agents must have knowledge of a social structure that describes their interrelationships. The agent modeler must design this structure carefully to capture these relationships. Social networks offer one such means both to describe and to visualize how each agent relates to the others.

Networks are composed of nodes and arcs. For agent-based models the nodes represent either individual agents or groups of agents, or external resources that somehow contribute to the model. The arcs define the relationships between pairs of nodes. They can represent any relationship envisioned by the modeler: for example, political, financial, or familial links. Figure 4-9 is an example of a three-node network. Note that arcs can be either one- or two-way. Two-way arcs indicate that the relationship goes both ways. Any number of relationships may exist between any pairs of nodes.

Arcs can also be viewed as paths for resource transfer. In Figure 4-9, node 1 may have a financial relationship with node 3, as depicted by their connecting arc. Assuming that a positive relationship exists along this arc, it could represent the positive flow of money from 1 to 3.

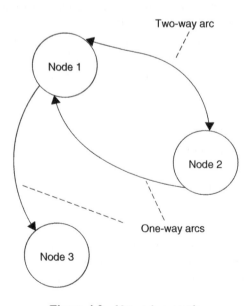

Figure 4-9   Network example.

Much scholarship exists that provide methods for identifying the underlying social networks of real-world systems and for analyzing these networks mathematically (*social network analysis*). In this book we use them as a tool for the modeler to lay out the relationships as part of a model design method that will facilitate implementation of the model in a working simulation. In the next section we illustrate this design method through the development of a simple agent-based model. It is also used in the case studies in Chapters 7 and 8.

BUILDING AN AGENT-BASED MODEL

A methodology for constructing an agent-based model is composed of seven steps:

1. Determine the purpose of the model and what type of output it will provide.
2. Determine the entities to be modeled by an agent.
3. Define the relationships among the agents and between each agent and the environment.
4. Build the social network structure that links agents through these relationships.
5. Determine the variables needed to represent each agent's behavior, including each agent's goals and its relative priorities.
6. Define and build the agent logic structure.
7. Validate the agent.

Models of global events are usually built for specific purposes, possibly to re-create the event or to explore alternative scenarios based on one point in time of the event's history. Recall from Chapter 2 that we can have predictive models that endeavor to answer specific questions. We can also have models that provide insight into the event, although they may not be valid for prediction purposes. It is important to understand the purpose of the model because that will define the scope of the modeling effort. Limiting the scope of the model to a precise purpose will prevent the modeler from constructing an unnecessarily complex model. Scoping the model will also help identify what entities to include and what behaviors will be required to fulfill the model's purpose.

When identifying the entities to include in the model, be careful to delineate between models that require agent representation and models that can exist as part of the external environment. Entities in the environment not represented by agents are those not requiring a behavior structure; they can usually be depicted as simple variables. All other entities should be represented by agents.

Once the foregoing step is complete, the modeler can construct a social network that links the agents together. This necessitates identifying what relationships among the agents are important for the model. It also goes hand in hand with the next step, which is what variables will be required to describe the agent and its behavior. Recall that the social network structure and the variables that characterize each agent are derived from steps 1 to 3 of the global event modeling methodology described in Chapter 2.

Choosing and implementing the agent logic structure will depend on the complexity of the agent behavior. The modeler may be able to employ a simple rule-based system. A complex logic structure consisting of a combination of those outlined in this chapter may be required when trying to represent actual human thought. (For an example of a multilogic structure system, review Sokolowski's discussion of the RPDAgent model [5].)

Recall the discussion of model validation in Chapter 2. It is important to include this step in the agent-based model as it ensures that the model is meeting its intended purpose.

CONCLUSIONS

This chapter described the key concepts behind agent-based modeling and social networks. It provided a definition for an agent, which pointed out the unique nature of this modeling paradigm. It also reviewed the major techniques employed by agent modelers to create behaviors for these entities. Agent communication and agent negotiation were also discussed as concepts that must be implemented to have a working agent system. Finally, social networks were discussed in the context of how they support the design of agent models, and a methodology was provided as a guideline for building agent-based models.

KEY TERMS

social network   reasoning process
agents   finite-state machine
autonomy   Markov chain
environment   rule-based model
goal   expert system
state-space model   knowledge engineering

forward chaining   fuzzy logic
backward chaining   fuzzy set
case-based reasoning   fuzzy logic
artificial neural network   fuzzy inference
back propagation

REFERENCES

[1] Shortliffe EH, Axline SG, Buchanan BG, Merigan TC, Cohen SN. An artificial intelligence program to advise physicians regarding antimicrobial therapy. *Computers in Biomedical Research* 1973;6:544–560.

[2] Dubois D, Prade H. *Fuzzy Sets and Systems*. New York: Academic Press, 1988.

[3] Wooldridge M. *An Introduction to MultiAgent Systems*. Chichester, UK: Wiley, 2002.

[4] FIPA. Specification part 2, *Agent Communication Language*. Apr. 16, 1999.

[5] Sokolowski JA. Enhanced decision modeling using multiagent system simulation. *Simulation* 2003;79(4):232–242.

5 Game Theory

INTRODUCTION

Game theory is a branch of applied mathematics used in numerous disciplines. In the social sciences game theory is used to capture human behavior that governs strategic decisions when a person's choice depends on and is affected by the choices of others. The origins of game theory can be found in a 1944 publication, *The Theory of Games and Economic Behavior*, by John von Neumann and Oskar Morgenstern [1]. Although this book focused on economic situations, it provided the foundation for using the concept in many areas. For example, a recent book by Steven J. Brams uses game theory to explore presidential election strategies from the primary process through the final election [2].

Games analyzed by game theory have well-defined mathematical representations. These games typically have two or more players who utilize a set of strategies to maximize their payoff or return. Through simulation of the game, players can investigate multiple strategies to determine which ones offer the greatest return or advantage. This chapter introduces various game-theoretic concepts and explore how these concepts can be combined to study problems and strategies as part of the analysis of global events.

FUNDAMENTALS OF GAME THEORY

To understand game theory we need a definition to provide a common reference. First, the definition of **game**: any social situation or interaction

Modeling and Simulation for Analyzing Global Events, By John A. Sokolowski and Catherine M. Banks

involving two or more people, called **players**. There exists two fundamental assumptions about players in these games: They are *rational* and they are *intelligent*. **Rational** implies that the players make consistent decisions that serve to advance their objectives or goals. In much of the game theory literature, especially that covering economics, rationality implies maximizing one's return on investment by choosing the highest monetary payoff calculated from a probability standpoint. For purposes of this discussion, rational behavior will not be defined so formally since many people have different concepts of gain. People (players) come from different cultures, which bring varying goals that each player may seek, yet these people still view themselves as rational actors. They may, however, be motivated by different factors. This behavior is explained by **utility theory**, where each player assigns a personal utility value to various possible outcomes and chooses the outcome that maximizes his or her expected return. *Return* is defined as the product of the utility value and the probability of the specific outcome.

In many cases these probabilities are not specific numbers; rather, they are subjective in nature. While a player may be able to assign a subjective probability to a specific choice, he must also evaluate the subjective probability choices of his opponent. Here he must place himself in his opponent's shoes and estimate how his opponent will specify his subjective evaluation of probability. So the rational solution or decision of both players depends on the other player's rational choice. Thus, their decision problems may be analyzed together. Game theory facilitates this type of analysis.

The term **intelligent** refers to a player's knowing everything about the game that the game designer does. For example, if the game designer believes that there is a specific set of policies that apply to the game, the player knows these policies and knows how to apply them in the same manner as the designer. Each player also knows the complete set of rules for the game and knows all other players' utility functions and the resulting payoffs.

Forms of Game Theory

Games take on several forms in game theory. The two most common are the *extensive form* and the *normal* or *strategic form*. The extensive form is the more structured of the two representations; the normal form is more conducive to general analysis.

Extensive Form   Games in extensive form are represented by tree structures, with each node of the tree either a decision point for a specific player or a chance node, and each edge of the tree representing a choice. The extensive form accounts for the following elements:

1. The players of a game
2. For every player, every opportunity that he has to move
3. What each player can do at each of her moves
4. What each player knows for each move
5. The payoffs received by every player for every possible combination of moves

Play begins at an initial node and progresses across the tree until a terminal node is reached. Here play ends and payoffs are assigned to each player. In contrast to the normal form, the extensive form is able to represent the complete sequence of moves from beginning to end. To illustrate this form, consider the game tree of Figure 5-1. Each node of the tree represents a decision point for one of the players. In Figure 5-1, player 1 must decide whether to choose path a or a' or b or b'. Then it is player 2's turn. Player 2 must make decision c or c' or d or d', but his decision depends on player 1's choice. Player 1's choice depends on what he thinks player 2 will do. In the end, each player is trying to maximize his payoff at the completion of the game.

One other key piece of information must be depicted with the game tree: What knowledge does a player have for a given branch of the tree? This knowledge is indicated by enclosing equivalent nodes by a dashed line. In Figure 5-1, player 2 does not know what choice player 1 has

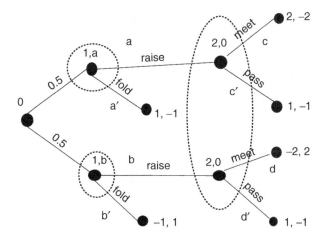

Figure 5-1 Extensive form game model.

made, so two nodes are encompassed representing this uncertainty and the choices facing player 2. Player 2 will make choice c or c' or d or d', but does not know at which branch he is. Player 1 has a single node circled. This represents player 1 as knowing exactly where he is on the game tree.

Normal (Strategic) Form   The normal (strategic) form is a somewhat simpler way of representing a game structure and is the one most often encountered in the literature when describing a specific game setup. Three items are necessary to specify a game in this form:

1. The set of players in the game
2. The options available to each player
3. The way a player's payoff depends on the option the chooses

The main difference between this form and the extensive form is that the normal form does not portray a sense of timing. Because the normal form assumes that all players make their choice of strategy simultaneously, it can only be used when questions of timing are not relevant.

Games in normal form are represented by matrices that have dimensions equal to the number of players in the game and the number of choices available to each player. The most common form is a two-person game with a $m \times n$ matrix, where m is the number of choices available to player 1 and n is the number of choices for player 2. The pair of numbers in each cell of the matrix are the payoff values for each player. The number on the left is player 1's payoff; the number on the right is player 2's payoff. This setup is depicted in Table 5-1.

These game forms are used to illustrate the various types of games found throughout game theory.

Game Characteristics

Next, we describe the more common game characteristics found throughout the literature of game theory.

TABLE 5-1   Normal Form Game Representation

		Player 2	
		c	d
Player 1	A	1,5	3,2
	B	4,0	1,0

Cooperative Versus Non-cooperative Games   **Cooperative games** are games in which players form coalitions. Each player brings a particular expertise or talent to the coalition with the idea of maximizing the return for the coalition as a whole. **Non-cooperative games** are concerned with the analysis of strategic choices. Here, individual player strategies are explored from the aspects of timing and ordering. In this type of game, players make choices based on their own self-interest and not on that of a group. This is not to say that there is not cooperation, which often does arise in non-cooperative games. The difference is whether or not the cooperation is self-motivated.

Symmetric Versus Asymmetric Games   **Symmetric games** are ones whose payoffs depend only on the strategies being played and not on the particular player playing them. In other words, players could switch positions and the game would not change except for that switch. **Asymmetric games** have different outcomes based on the particular player. Table 5-2 is a normal form payoff matrix illustrating a symmetric game. Table 5-1 represents an asymmetric payoff matrix.

Zero-Sum Versus Non-Zero-Sum Games   A **zero-sum game** is one in which the total benefit to all players sums to zero. Choices by players can neither increase nor decrease the resources available. Sporting events where there are both winning and losing teams are examples of zero-sum games. Conversely, **non-zero-sum games** have total payoffs that add to greater or less than zero. It is possible for each player to gain by the policy he decides to pursue. Table 5-3 is a representative zero-sum game, while Table 5-1 depicts a non-zero-sum payoff matrix.

TABLE 5-2   Symmetric Payoff Matrix

		Player 2 c	Player 2 d
	A	1,2	0,0
Player 1	B	0,0	1,2

TABLE 5-3   Zero-Sum Payoff Matrix Example

		Player 2 c	Player 2 d
	A	−1, 1	3, −3
Player 1	B	0,0	−2, 2

Simultaneous Versus Sequential Games   A **simultaneous game** is one in which both players move at the same time; or if they don't move at the same time, they are unaware of how the other players have moved. This scenario essentially makes a game simultaneous. **Sequential games** involve a series of moves, with the next move predicated based on the one before. Simultaneous games are represented using the normal form (see Table 5-1), whereas sequential games must use the extensive form (see Figure 5-1) for their representation.

Perfect Information Versus Imperfect Information Games   Perfect versus imperfect information games are a subset of sequential games. In a perfect information game each player knows the previous moves of all other players and thus can make a decision fully aware of how the other players have moved up to that point. Most games do not involve perfect information, but there are some commonly known examples of games that do portray perfect knowledge, such as chess and checkers.

TYPES OF GAMES

As game theory evolved, many types of games were developed to capture the concepts they were meant to represent. Discussed below are the most common types of these games, together with examples that explore their characteristics.

Two-Person Zero-Sum Games

This study of types of game theory begins with two-person zero-sum games, which, as the name implies, consist of two players competing with one another in a win–lose outcome. Players must choose their strategies simultaneously. This type of game is usually represented by the normal form. Table 5-4 represents such a game, with player 1 having choices A, B, or C and player 2 having choices I, II, or III. The numbers in the cells represent how much player 2 has to pay player 1, with negative numbers

TABLE 5-4   Two-Person Zero-Sum Game Example

		Player 2		
		I	II	III
	A	5	−2	1
Player 1	B	6	4	2
	C	0	7	−1

indicating what player 2 is paid. This notation is slightly different from that presented in Table 5-3, in that player 2's values are left off and assumed to be the negative of player 1's values. It is zero-sum because one person wins the amount in the cell and the other person loses that amount.

So if player 1 chooses A and player 2 chooses II, player 2 is paid $2 by player 1. What is the best strategy for each player to choose? In the choice above, player 1 loses $2—that does not appear to be the best choice. If player 2 thinks player 1 will choose C, he will choose III, and again player 1 loses money. If B is chosen by player 1, player 2 will choose III. Here player 1 wins $2 and player 2 minimizes his losses over choosing I or II.

Therefore, given that each player must simultaneously declare his or her strategy, B and III appear to be the best pair to choose since each player maximizes gain or minimizes loss. Neither player can do any better than this given the fact that each player has full knowledge of the game, its payoffs and its rules. These two strategies are called **equilibrium strategies**, and the payoff associated with these strategies is called an equilibrium point or a *Nash equilibrium*, after John Nash, winner of a Nobel prize for his work in this area [3]. Once equilibrium strategies are found, the players have no rational reason to deviate from them. If one of these points does exist, it is referred to as a *solution to the game*.

In two-person zero-sum games, equilibrium points, if they exist, are easy to find. The payoff at an equilibrium point is always the *largest value in its column and the smallest value in its row*. To find these points, examine each column for its maximum value and see if that value is also the smallest in its row. Do this for each column to find all the points, keeping in mind that an equilibrium point may not exist. Are there any other equilibrium points in Table 5-4?

Game theory also engages the concept of *domination*. Simply, one strategy is dominant over another if that strategy produces payoffs that are better than or at least as good as another strategy. Consider the game in Table 5-5. Here, strategy B dominates strategies A and C since it has higher payoffs in all cases. Strategies A and C are said to be *dominated*

TABLE 5-5   Dominant Strategy Illustration

		Player 2		
		I	II	III
Player 1	A	7	9	8
	B	9	10	12
	C	8	8	8

by strategy B. Strategy I dominates strategies II and III because player 2 has the least losses for this strategy. When analyzing a zero-sum game, one should assume the following:

1. You will never pick a *dominated strategy*. You can always do better.
2. Your opponent will never play a *dominated strategy*. He can always do better.

In some cases a dominant strategy may not be readily apparent for one of the players. To see this, examine the game in Table 5-6. There is no dominant strategy for player 1. For player 2, strategy III dominates strategy I, so I can be eliminated. With I eliminated, player 1 sees that strategy B dominates strategies A and C, so these can be eliminated. With these gone, player 2 sees that III dominates II and an equilibrium point exists with cell value 3.

Thus far we have examined two-person zero-sum games that have equilibrium points. But how do we play if one or more of these points do not exist? Take, for example, a coin-flipping game with the payoff matrix given in Table 5-7. In this game player 1 receives the payoffs indicated in the table. If he chooses heads, player 2 will choose heads, and vice versa. So there does not appear to be any strategy in which player 1 can avoid losing \$1. However, this is not the case, as noted in the following analysis.

The game shown in Table 5-7 depicts playing *pure strategy*. It is the same strategy used repeatedly and does not afford the player an opportunity to minimize his losses. However, using probability in his favor does

TABLE 5-6 Finding Dominant Strategies

		Player 2		
		I	II	III
Player 1	A	19	0	1
	B	11	9	3
	C	23	7	−3

TABLE 5-7 Two-Person Zero-Sum Game with No Equilibrium Points

		Player 2	
		Heads	Tails
Player 1	Heads	−1	+1
	Tails	+1	−1

affect the outcome of this game. By using a chance device such as a die to choose a strategy allows the player to break even at this game. This type of strategy is known as a **mixed strategy**.

Using a chance device affords the opportunity to play an infinite number of mixed strategies. For example, assume that player 1 chooses heads with probability p and tails with probability $1 - p$ with p being a number between zero and one. Further, assume that he plays heads half the time ($p = 0.5$) and that the selection of the play is controlled by rolling a fair die, where heads are played when the die roll is 1, 2, or 3 and tails are played for die roll 4, 5, or 6. Player 2 cannot predict the outcome of each roll of the die, so the best he can do is also choose his strategy randomly. This means that, on average, each player will win half the time and lose half the time, thus breaking even over the long run. This process is clearly better than playing a pure heads or tails strategy all the time.

The foregoing outcome is described by von Neumann in his **mini-max theorem**, which states that one can assign to every finite two-person zero-sum game a value V: the average amount that player 1 can expect to win from player 2 if both players play rationally [1]. V is known as the *value* of the game. This is one of the most important theorems in game theory because of its contribution to solving these types of games.

Theoretically, it is possible to determine V for all finite two-person zero-sum games, but it may be difficult. The following simple method provides one means for solving some of these types of games, but it does not work in all situations. The procedure is as follows.

1. Use the procedure outlined above to determine if Nash equilibria exist. If not, proceed to step 2.
2. Eliminate all dominated strategies.
3. Assign probabilities to the remaining strategies and to those of the opponent.
4. Calculate these probabilities and determine player 1 and his opponent's outcomes. If the outcomes are the same and are nonnegative, the game is solved. If not, this procedure will not work.

As an example of this procedure, consider the game depicted in Table 5-8. Following the procedure outlined above, one would first determine if there were any equilibrium solutions. In this case, none exists. Next, one would eliminate any dominated strategies. No dominated strategies exist. Finally, there is the assigning of probabilities to the remaining strategies for each player and the calculation of the value of the game for each. Let p represent the probability of playing strategy

TABLE 5-8   **Calculation of Mixed Strategy Value**

		Player 2		
		I	II	III
	A	7	5	13
Player 1	B	5	14	2
	C	17	5	5

A, q the probability of B, and $(1 - p - q)$ the probability for C. For player 2, let r represent the probability for playing strategy I, s the probability for II, and $(1 - r - s)$ the probability for III. Then the equation for player 1 is

$$7p + 5q + 17(1 - p - q) = 5p + 14q + 5(1 - p - q)$$
$$= 13p + 2q + 5(1 - p - q) \qquad (5\text{-}1)$$

and the equation for player 2 is

$$7r + 5s + 13(1 - r - s) = 5r + 14s + 2(1 - r - s)$$
$$= 17r + 5s + 5(1 - r - s) \qquad (5\text{-}2)$$

Solving equation (5-1) for p and q yields $p = \frac{1}{2}, q = \frac{1}{3}$, and $1 - p - q = \frac{1}{6}$. So strategy A is played one-half of the time, B one-third of the time, and C one-sixth of the time, giving a value of 8 in the game for player 1.

For player 2, solving equation (5-2) for r and s yields $r = \frac{1}{4}, s = \frac{7}{16}$, and $(1 - r - s) = \frac{5}{16}$. So strategy I is played one-fourth of the time, II seven-sixteenths of the time, and III five-sixteenths of the time, providing a value of 8 in the game for player 2. So this game was solvable by the method described above.

In the next section we explore two-person non-zero-sum games where both players have an opportunity to benefit from their combined strategies.

Two-Person Non-Zero-Sum Games

Two-person non-zero-sum games are discussed in the context of the classic game theory example, the Prisoner's Dilemma.

Prisoner's Dilemma   The game Prisoner's Dilemma is a classic example used to illustrate various game theory concepts. In this game there are two players who have committed a crime for which there is only circumstantial evidence. The players are put into separate jail cells and not allowed to communicate with one another. The police try to pit one player against the other by giving each the choice of confessing to the crime or denying his involvement. The prison sentence that each receives depends on the other's choice. If player 1 confesses and player 2 denies involvement, player 2 receives a 10-year sentence. The same is true if player 2 confesses and player 1 denies. If they both confess, each receives a five-year sentence, and if they both deny, they each receive a one-year sentence on circumstantial evidence.

The Prisoner's Dilemma game is usually represented in the normal form since both prisoners must make a decision without being able to communicate with the another. Essentially, they make their decisions simultaneously. The normal representation for this game is shown in Table 5-9. The game can be classified as (1) a two-person non-zero-sum game, (2) a simultaneous game, and (3) a non-cooperative game. It is non-zero-sum because each player can benefit if the proper choice is made. It is simultaneous and non-cooperative because each player must make a decision without knowing the other's move.

We will analyze the game from each player's perspective. Player 1 does not know if player 2 will confess or deny, but he wants to minimize his punishment. He must consider two cases.

1. If player 2 confesses, player 1 will receive five years if he confesses or 10 years if he denies. So it is better for him to confess.
2. If player 2 denies, player 1 will go free if he confesses or receive one year if he denies. So again, it is better to confess.

A similar analysis applies for player 2.

1. If player 1 confesses, player 2 will receive five years if she confesses and 10 years if she denies.
2. If player 1 denies, player 2 will go free if she confesses or receive one year if she denies. It is better for player 2 to confess.

TABLE 5-9 Prisoner's Dilemma Game Normal Representation

		Player 2	
		Confess	Deny
Player 1	Confess	5,5	0,10
	Deny	10,0	1,1

So both players gain if they both confess; hence the non-zero-sum aspect of the game.

From player 1's point of view, confess dominates deny because player 1 always receives a lesser punishment when he confesses regardless of whether player 2 confesses or denies. The same holds true for player 2.

The Prisoner's Dilemma example above has one Nash equilibrium. As shown previously, it is the strategy of *confess, confess*. For non-zero-sum games there is a straightforward way of identifying Nash equilibria in a payoff matrix. The rule goes as follows. If the first payoff number in a cell is the maximum for its column and the second payoff number in the cell is the maximum for its row, then the cell contains a Nash equilibrium. Table 5-10 illustrates this concept. The payoff pairs in bold are Nash equilibria as identified by this process.

The Prisoner's Dilemma example does not make any assumption about the likelihood of player 1 or player 2 confessing. If we introduce the concept of probability and assigns to each player a probability of confessing versus not confessing, a formula can be developed to represent the expected payoff based on these likelihoods. This is another example of playing a mixed strategy. Assume that player 1 confesses with a probability q and player 1 assumes that player 2 will confess with a probability p. Then the payoff expected for player 1 can be calculated as follows:

$$p \times q \times 5 + p \times q \times 10 + p \times q \times 0 + p \times q \times 1 = 16pq \qquad (5\text{-}3)$$

Equation (5-3) is a decreasing function in q, so the more likely that player 1 is to confess, the less punishment he will receive, irrespective of player 2's choice. A similar equation can be developed for player 2, with the same results. The same process for computing mixed-strategy outcomes as was followed for zero-sum games can also be followed for non-zero-sum games. But again, there is no guarantee of finding a solution.

How would the outcome of the Prisoner's Dilemma game change if the players were allowed to cooperate? From the payoff matrix of Table 5-9, if both denied involvement, they would each receive a one-year sentence. This is clearly preferable to both confessing. But this choice depends on

TABLE 5-10   Nash Equilibrium Determination

		Player 2		
		Option A	Option B	Option C
	Option 1	10,0	**25,40**	5,10
Player 1	Option 2	**40,25**	0,0	5,15
	Option 3	10,5	15,5	**10,10**

how much one player trusts the other. If they both agree on denying and then one player decides to break the commitment, he or she would go free. What are the consequences if one of the players decides to break the agreement? The safe and rational choice is to stick with the Nash equilibrium. However, in the real-world people face these types of situations all the time. Understanding how to apply game-theoretic concepts provides us with a valuable tool to investigate the best strategy.

N-Person Games

N-person games are a generalization of two-person games. In two-person zero-sum games with equilibrium points, each player could control his own destiny. Even without equilibrium points, players in this type of game could arrive at a mixed strategy that guaranteed equilibrium. In two-person non-zero-sum games, each player's fate was shared with the other player, but he also controlled the fate of his opponent. In n-person games, generally none of the above is true. For players to maximize their outcome they must form coalitions for mutual benefit, much like the cooperation option pointed out in the Prisoner's Dilemma example. So the issue becomes one of power. Specifically, the power that an individual player or a coalition of players possesses to affect the outcome of the game is of primary concern in n-person games. However, the concept of power is somewhat elusive in an n-person game. There is always a minimum payoff that a player can receive by himself. To get more, he must form a coalition. He also brings some value to the coalition, so he has *potential* power that can be realized if cooperation is achieved.

N-person games are very complex and do not lend themselves to the same analysis methods as those found in two-person games. However, von Neumann and Morgenstern did propose a theory on how to approach this problem. To explain their theory, here called *N-M theory*, assume that each player in an n-person game can achieve a certain payoff if he acts by himself. Then suppose that two or more players form a coalition S. What is the value of forming the coalition? N-M theory states that this situation can be reduced to a two-person zero-sum game, with player 1 being the coalition of players (S) and player 2 being everyone else. Then $V(S)$ is the value of the coalition and can be calculated as discussed previously. This type of game is said to be in *characteristic function form*, where every coalition has a specified value. There must also be a specified rule on how the value of combining coalitions must be treated. Suppose that there are two coalitions, R and S, that have no members in common. Now a new coalition is formed that includes all the members of R or S ($R \cup S$ is read "R union S"). Clearly, the value of this union must be at least as great as the sum of the individual coalition values, so, mathematically, $V(R) + V(S) \leq V(R \cup S)$. This concept is known as *superadditivity*.

The problem with N-M theory occurs when more than one coalition is formed. At that point it ceases to be a two-person game, and other techniques must be followed to look for a solution. However, the characteristic function form still provides a method to study the problem. To illustrate how this approach is used, consider the following problem. Three neighboring countries (A, B, and C) have natural resources that can be used to enhance their economic development. The value of each country's resources is the same and is set at 5. Any two of them or all three can form a coalition. If such a coalition is formed, the combined resources of each country provides an extra 6 units of value above the 5 units that each country enjoys already. In that case, a two-country coalition would be worth 16 units $(5 + 5 + 6)$. A three-country coalition would have a value of 21 units. In characteristic function form this would be written as $V(i) = 5, i = $ A,B,C, $V(AB) = V(BC) = V(AC) = 16$, and $V(ABC) = 21$. So each country acting alone would have an economic development worth of 5, which will be indicated by the tuple (5, 5, 5). If A and B formed a coalition and divided the payoff equally, each would receive a value of 8, and C would have only 5 (8, 8, 5). For the coalition of three members, dividing the value equally, each would receive seven (7, 7, 7). So what is the best strategy for each country?

Suppose that country C proposes no coalition and that each country uses its resources individually. The payoff would be (5, 5, 5). But country A recognizes that it can do better and forms a coalition with B for a payoff of (8, 8, 5). Country C appears to be left out of the deal. But suppose that country C makes a counteroffer to country B, saying that it would split the value of the coalition, giving country B 9 and country C taking 7 with the resulting payoff of (5, 9, 7). Clearly, countries B and C are better off and A is left out of the deal. This type of bargaining could go on endlessly, and no formal theory has been developed to guide the selection of a proper strategy. However, some guidelines exist to reduce the complexity of the problem.

One approach is to reduce the number of possible payoffs by removing those that would not be selected rationally. Option (5, 5, 5) would not be chosen when all players (countries) could get (7, 7, 7). N-M theory assumes that the ultimate payoff will be *Pareto optimal*, which means that there is no payoff where all players (countries) could do better simultaneously. N-M theory also assumes that each player (country) will only accept a solution in which he gets at least as much as if he had gone it alone. This is known as *individual rationality*. So a solution of (5, 12, 4) would not be individually rational.

Assuming that only Pareto optimal, individually rational solutions are considered, if a proposal is made, under what conditions would an alternative proposal be acceptable and pursued? Two conditions would have to exist for the alternative to go forward: (1) the new coalition must be strong enough to go forward, and (2) the members of the new coalition must get more than they would have if they had stayed with the old proposal. If this is the case, the new proposal *dominates* the old one. To illustrate, assume that the original coalition payoff was (8, 7, 6). Now suppose that B and C propose (6, 8, 7). This is certainly preferred by B and C since they each get an extra unit of value. So (6, 8, 7) dominates (8, 7, 6) and becomes what is known as the *effective set*. What if (1, 17, 3) were proposed? In this case no two players would support this proposal since only B benefits but does not have the power to enforce this option.

In reality, there may be many undominated solutions or there may be no undominated solutions. So N-M theory does not purport to have a method to solve all *n*-person games; however, it does provide an approach to analyzing them. And in some cases a solution to the game may not exist, as was shown by Lucas [4]. Other methods have been developed to address *n*-person games, such as the *Aumann–Maschler theory* [5] and the *shapely value* [6]. (The reader is encouraged to explore these methods and to compare them to the N-M approach.)

CONCLUSIONS

As one can see, game theory provides a reasoned approach to modeling decision-making scenarios. It offers a method for thinking about the advantages of certain policies in both a competitive and a cooperative environment, which helps shed light on selecting the optimal choice on which to proceed. It transcends the purely academic pursuit of analyzing problems by showing that better decisions can be made in any area where results can be quantified.

KEY TERMS

game theory   intelligent
game   cooperative games
players   non-cooperative games
rational   symmetric games
utility theory   asymmetric games

zero-sum games

non-zero-sum games

simultaneous games

sequential games

two-person games

equilibrium strategies

mixed strategy

minimax theorem

n-person games

REFERENCES

[1] von Neumann J, Morgenstern O. *The Theory of Games and Economic Behavior*, 3rd ed. Princeton; NJ: Princeton University Press, 1953.

[2] Brams SJ. *The Presidential Election Game*, rev. ed. Wellesley, A K Peters, Wellesley, MA: 2008.

[3] Nash J. Equilibrium points in n-person games. *Proceedings of the National Academy of Sciences* 1950;36(1):48–49.

[4] Lucas WF. A game with no solution. *Bulletin of the American Mathematical Society* 1968;74:237–239.

[5] Aumann RJ, Maschler M. The bargaining set for cooperative games. In: Dreshner M, Shapely LS, Tucker AW, eds. *Advances in Game Theory*. Annals of Mathematical Study 52. Princeton, NJ: Princeton University Press, 1964, pp. 443–476.

[6] Shapely LS. A value for n-person games. In: Kuhn HW, Tucker AW, eds. *Contributions to Game Theory*. Princeton, NJ: Princeton University Press, 1953, pp. 307–317.

PART III
Modeling Global Events

6 Case Study: Colombia—A Country Study of Insurgency

INTRODUCTION

Understanding the nature of conflict in the twenty-first century is critical in determining strategies for intervention. What compels one's attention is the fact that conflict can now transcend national and/or state borders, engage totally in nontraditional military tactics, and preclude diplomatic resources. Traditional warfare objectives, counterinsurgency strategy, and nation-building methodologies are undergoing revision to integrate historical, cultural, and political awareness into its decision-making capacity. Institutions must react and respond to disturbances that are the result of insurgencies, military action, terrorist attack, and natural disaster. To engage proactively it is necessary to understand the *what happened* and to explore the *what if*. The challenge for government agencies and non-governmental organizations is often twofold: defining the dilemma and deciding on a resolution.

Recent attempts at understanding and explaining the behavior of insurgent or terrorist networks have failed for two reasons: (1) the models focused on the behavior of one as the baseline for predicting the behavior of the whole, and (2) the models were developed on the premise that insurgency is chaotic and that defeating insurgency requires removing the leader.

An insurgent network is more than an interconnection of acts. At a minimum, these networks encompass many things, including acts of terrorism, relationships among actors, philosophy, social structure, organization and unity of the network, strategy, geography, external support (financial, moral, political), communication (internal and external), weapons, and

Modeling and Simulation for Analyzing Global Events, By John A. Sokolowski and Catherine M. Banks

access or opportunity (means). Insurgency and acts of terrorism have become for many a means of achieving a goal, a tactic. Hence, analyzing insurgent networks is much more than connecting the dots and eliminating the leader because that understates the difficulty of making these connections, due to the abundance of data mining necessary and sorting through the mis-, dis-, and outdated information. When analyzing insurgencies one must recognize that networks do not capture the dynamic relationships of all the variables that define an insurgency.

History is replete with examples of insurgency and acts of terrorism which prove that the behavior of one cannot typify the behavior of a group, or that the loss of leadership can create even greater chaos. This became apparent with the death of al-Qaeda terrorist leader Abu Musab al-Zarqawi in Iraq in June 2006. Although thought to be a *good omen* by U.S. ambassador in Iraq Zalmay Khalilzad, Zarqawi's death did not result in the end of the insurgent attacks nor of the insurgent network in Iraq. To compartmentalize insurgents and their networks is to say that all insurgent activities are motivated for the same reasons and executed in the same manner. Neither is true. Motivations vary from insurgency to insurgency and within the insurgent network from member to member. Execution of these nefarious acts can differ with that same degree of variation.

In the country of Colombia, the collapse of the two major drug cartels did not result in the end of the drug trade and the **Dirty War** associated with it.[1] The death of Pablo Escobar in 1993 may have brought an end to his leadership of the Medellin cartel, but it resulted in a chaotic frenzy of unstructured, unorganized drug trafficking that no longer adhered to the cartel's *espirit Mafioso*.[2] Moreover, the loss of leadership democratized the drug trade, facilitating the splintering and multiplying of many smaller cartels. Additionally, what was once a *War on Drugs* morphed into the *Global War on Terrorism* when Presidents George W. Bush and Alvaro

---

[1]The term Dirty War is derived from the disappearances, executions, assassinations, detentions, repression, and massacres of civilians.

[2]*Espirit Mafioso*, the Mafia, is not defined as a formal organization but as a form of behavior and a mode of power (M. Matard-Bonucci, *The History of the Mafia*. Brussels, Belgium: Editions Complexe, 1994). There are four reasons why a Mafia can come into existence:

1. The state does not have a firm presence and local forces contest exclusionary authority.
2. There is personalized mediation with centralized political powers—an elite representing the people.
3. Dependency is mediated through the use of economic resources and political power.
4. Private violence is used as a mechanism of social control.

Uribe decided to fund and conduct the counterinsurgency against the drug traffickers as part of the larger war on terrorism.[3]

In this chapter we introduce modeling insurgencies as an acceptable means to gain insight into the various characteristics of asymmetric warfare to proffer prescriptive resolutions for mitigating their effects. The Colombian insurgency poses the challenge of assessing political behavior in a nontraditional revolutionary climate. Factors prevalent in traditional insurgency are not applicable in Colombia, specifically between the years 1993 and 2001 with the **democratization of the drug trade**. This was followed by the catastrophic events of September 11, 2001, which reverberated in Colombia, resulting in a new policy and strategy in waging the counterinsurgency there. This case study introduces four things: (1) a structured methodology to modeling the Colombian counterinsurgency, incorporating qualitative assessment to provide a historical analysis of Colombia's internal disorder and counterinsurgency, (2) a discussion on measuring the capacity of the counterinsurgency and the translation of those qualitative findings into quantifiable data, (3) a concise explanation of the system dynamics modeling method for assessing the counterinsurgency, and (4) a response to the research question.

DEVELOPING THE RESEARCH QUESTION AND METHODOLOGY

This case study is tasked with characterizing the marginal and long-term changes in population numbers and insurgency strength due to modifications in governmental policy and military strategy for the counterinsurgency. The research centers on the effects or outcome of executing new government policy in waging a deeply entrenched and seemingly unending insurgency. Therefore, the research question seeks to answer: *How can governmental policy changes in Colombia's War on Drugs be measured, then represented in a qualitatively developed and quantifiably supported model that can predict insurgency strength and prescribe counterinsurgency strategy?*

Integral to this case study is a review of many aspects of society (history, culture, religion, government, economics, security) that affect the behavior of the civilian population in the support or nonsupport of the insurgency as a way of explaining the difficult problem of conducting a

---

[3] Subsuming the Colombian insurgency and its War on Drugs into the Global War on Terrorism has also resulted in the arbitrariness of the definition of insurgency; that is, *revolution, guerrilla warfare, asymmetric warfare*, and *terrorism* are often used interchangeably.

counterinsurgency. The research then engages M&S as a tool in developing a mathematical formulation of a conceptual model that includes a body of empirical (observable) data to substantiate the formulation. The goal is to develop a mathematical formula derived from a qualitative assessment that will lead to a predictive model in assessing the possible behavioral outcomes of the insurgency. (An added benefit would be to use the mathematical model to prescribe and proscribe counterinsurgent strategies.)

With the research question in hand, a recommended methodology for this study includes four basic steps:

1. Research the social sciences literature to provide a historical context and an assessment of Colombia's counterinsurgency activities.
2. Develop a means of measuring the insurgency and counterinsurgency capacity as well as other major variables that affect all segments of the population (insurgents, civilians, all members involved in the counterinsurgency effort).
3. Select a modeling method—in this case study it is system dynamics.
4. Apply a system dynamics model to validate the findings and present the results of the study.

The discussion below is based on qualitative research from primary and secondary sources on the subject of Colombia's insurgency. A bibliography for this case study appears at the end of the chapter.

BACKGROUND: QUALITATIVE RESEARCH

There is no shortage of literature assessing the events of Colombia's War on Drugs during the 1980s and 1990s and its current War on Terrorism. Within that body of research there is general agreement that the introduction of drugs served as the catalyst that shifted the intent of Colombia's insurgency. Colombia had experienced traditional (nationalistic) insurgency in its fight for independence against Spain and then suffered the woes of internal commotion and insurgency aimed at its own government as Colombians sought control of the country's political and social destiny. These clashes were about empowering the people and establishing a government that would provide services as well as secure the population. This was never realized in Colombia because drug trafficking took the country by storm. The drug trade marginalized the people. Members of the government became complicit and corrupt. All too often the government yielded because it had confronted the drug trafficking as it evolved into an insurgency problem in a weakened state. What led to this weakened condition, and why has the Colombian government been so ineffective?

Like many states in Central and South America, Colombia has experienced the spectrum of ills that come with being a former possession of a global power. With that discontent came a war for independence followed by long-term domestic discord over political ideals supported by two dominant parties: the Conservatives and the Liberals. Colombia was precluded by these conflicts from developing a sense of nationalism. Its government (led by both the Conservatives and the Liberals) was never able to unite its multi-ethnic population as *Colombians*. This internal commotion gave way to marginalization for political and economic power by those who saw a way to capitalize on Colombia's most lucrative export, coca.

The effects of the sale of coca on the state of Colombia are profound. The trafficking of the drug introduced a form of terrorism that brought with it paramilitaries, private armies, corruption, human rights abuses, internally displaced peoples, and a Dirty War against many innocent civilians. The Colombian counterinsurgency was relatively ineffective until the United States decided that the American War on Drugs should be waged jointly with Colombia. With U.S. support and an ideological shift that transitioned the War on Drugs into the Global War on Terrorism, the number of guerrillas, insurgents, and traffickers decreased. Let's see how this happened.

First we review Colombia's modern history, the formation of its two major political parties, and the events that gave rise to the dissident groups that later evolved into drug cartels and narco-terrorists. Then we discuss the period 1993–2001 to explain the consequences of the War on Drugs that led to dissolution of the cartels. Finally, we examine the post-September 11 period from 2001 to 2006 when the War on Drugs was subsumed into the Global War on Terrorism.

The first Spanish settlement in the region of Colombia was established in 1510, with the cities of Cartagena and Bogotá founded in 1533 and 1538. By 1740 the territory had become known as the *Viceroyalty of Nueva Granada* and it extended throughout Panama, Venezuela, and Ecuador. Like its neighbors, Colombia sought independence from Spain. In 1810 the country removed its Spanish dominion and declared itself *Greater Colombia*. The war for independence was led by Simón Bolívar and Francisco Santander. Bolívar was elected the first president of Greater Colombia; Santander was his vice president. Followers of Bolívar established the Conservative Party, while followers of Santander established the Liberal Party. The new country claimed rights to the states of the former Viceroyalty: Panama, Venezuela, and Ecuador. By 1830, Venezuela and Ecuador seceded from Greater Colombia. The *Republic of New Granada* was formed from the remaining states of Colombia and Panama.

The Republic of New Granada was not without its tensions, as political and economic rivalry between the indigenous populations of the region resulted in civil wars and harsh dictatorships. The **Conservative Party** and its followers sought a centralized government, close ties to the Catholic church, and limited suffrage. The **Liberal Party** sought a decentralized government with home rule, separation of church and state in education and civil matters, and broad voting rights. As one can imagine, this resulted in ideological swings each time the opposing political party laid claim to the executive office. Instability persisted throughout the nineteenth century. This instability brought with it two periods in which the military seized power: the first, in 1830, after dissolution of Greater Colombia, the second, in 1854, marked by dissent and unrest between political factions. After the first two coups, civilian rule was restored within the first year. A third military seizure of power took place in 1953 and lasted until 1957.

This coup was at the request of the government as a way to end Colombia's second civil war, which began in 1948 with the assassination of Jorge Gaitan, a populist liberal who was the front-running candidate for president.[4] The scene was frantic, as an angry mob chased and killed the assassin. Rioting ensued, resulting in over 1500 deaths. The violence spread to the countryside, with over 200,000 Colombians escaping to the country to get away from the violence. 1950 brought with it a Conservative government that instituted repressive policies aimed at controlling the violence. Colombia's president proposed a new constitution in 1953; the military responded by removing him from office.

That third military coup (1953) overthrew the Conservative government and ended the period of **La Violencia**. The coup was led by General

[4]Colombia's first civil war, called the *Thousand Days War*, was fought primarily in Panama. The Liberals had planned a revolt against the Conservatives by creating a disturbance in the department (state) of Panama to divert the Conservative Army away from Colombia. After many skirmishes and a near defeat for the Liberal army, a peace treaty was signed on July 24, 1900. With the treaty the Liberal army was ordered to give up its arms. The Conservative Governor of Panama, Carlos Alban, ordered the capture of Liberal political leader Victor Lorenzo. This led to armed pursuit throughout the mountainous region of Panama, with villages burned and Indians killed. The Indians revolted against the Conservatives and a guerrilla war began. The war ensued until the Liberal army was forced to surrender a second time. Another treaty was signed on November 29, 1901. A final engagement, assisted by the U.S. Navy, allowed the Conservative army to recoup. They were able to put down the Panamanian (Indian) rebellion. The United States disrupted further retaliatory engagements on the part of the Conservative army by refusing to allow attacks on Panama City and the railroad that ran along the city. A final treaty to end the Thousand Days War was signed on November 19, 1902 on board the *U.S.S. Wisconsin*.

Gustavo Rojas Pinilla. He remained in power for four years. His refusal to restore democracy led to his demise, as the military broke from the General and took back the country with bipartisan political support. A provisional government called the **National Front** was instituted. It remained active for 116 years, from 1858 to 1974. The two parties agreed that this new government would be administered through alternate terms in the executive office by each political party. As each party took its turn in the chief executive role, it took full advantage of its four-year term in office. This type of exploitation often leads to packed courts, hand-picked government appointments, election tampering at the state and local levels, and limited third-party participation. Colombia suffered all of these political ills.

In 1964, internal dissent and unrest escalated as many Colombians sought a greater practice of democracy by the National Front, as it was not without political repression. Guerrilla insurgencies became organized movements in response to political corruption and election fraud. Insurgent activity took its toll on the National Front. Between 1974 and 1982 each elected government, alternating between the Conservatives and the Liberals, had to act to end the insurgencies. The government sought the support of the civilian population by claiming to represent it against the rich, the powerful, and the corrupt. One insurgent movement, the *M-19* (April 19th Movement), argued the same, in addition to demanding land reform for the people. During this time, guerrilla movements challenged the government's authority and legitimacy. These movements or insurgencies were of a more traditional nature, suing for the end of government corruption and political empowerment of the people.

In the same year the *Revolutionary Armed Forces of Colombia* (**FARC**) was established. Much has been written about this organization, and it still exists as Colombia's largest insurgent group. The FARC formed in opposition to the National Front as a response to the official violence and militarist aggression on the part of the government. FARC introduced agrarian programs that won the support of farmers while attracting Liberals and Communists. Its goal was to confront the Colombian army and its ally, the United States. The FARC conducted its own politics, with the masses providing security and introducing social programs. Their propaganda focused on making certain that the civilian population perceived them as undefeated [1].

It was not long before the *National Liberation Army* (**ELN**) emerged. This army was not a genuine Colombian peasant movement because its origins were in Cuba and it drew its base from the disaffected middle class. The ELN was much smaller than FARC and nearly dissipated in the 1970s. It then reestablished itself in the 1980s as a politico-military organization with much smaller numbers than FARC. The nature of the

ELN's insurgent activity was difficult for the Colombian government to control, as it became infamous for exploding pipelines as a way to confront the oil concessions given foreign oil companies [1].

In the 1990s the United Self-Defense Forces of Colombia (**AUC**) came into the fray. The AUC comprised right-wing paramilitaries supported by wealthy landowners, drug cartels, and segments of the Colombian military. These forces committed numerous human rights abuses, including assassinations of politicians, leftist guerrillas, activists, and civilians.

Since the official declaration of the War on Drugs by President Nixon in 1973, the United States provided much in the way of military support (training) and funding. Although this resulted in fluctuations in the number of insurgents, the next two decades experienced the escalation of wealth and power from drug trafficking and the eventual centralization of the drug trade by a few powerful drug lords. This centralization continued to evolve into two major organizations that controlled the Colombian drug trade: the Medellin cartel and the Cali cartel. With the introduction of crack cocaine in the 1980s, the floodgates of the drug trade burst open. Crack was cheaper, more addictive, smoked versus injected, and was now readily available for casual use by a much larger audience. President George H.W. Bush affirmed the need for continued U.S. involvement in Colombia's counterinsurgency. By 1989 crack cocaine was having a significant effect on U.S. national security policy.

The cartels thrived because they developed a mafia-like operating system with enough wealth, influence, and fearlessness. The cartels were the first to hire private militaries to protect property, operations, and individuals. In the 1990s the conflicts between and among Colombian government forces, anti-government insurgent groups, and illegal paramilitary groups (the latter two heavily funded by the drug trade) escalated. Fortunately for the Colombian government, the insurgents lacked the military and popular support necessary to overthrow the state and take control of Colombia. Then, in 1993, Pablo Escobar was assassinated. His death, coupled with heavy U.S. support of the Colombian counterinsurgency, brought about the eventual demise of the two major cartels, which resulted in the democratization of the drug trade [2]. This sharing of opportunity spawned numerous insurgent groups, as many as 82 smaller cartels, and it increased the numbers of insurgents [3]. Colombia's counterinsurgency was underfunded, unprepared, ill-equipped, and suffering war fatigue. In 1993, President William Clinton committed U.S. support to a new program, **Plan Colombia**. The plan provided $1.3 billion to resolve the 34-year internal conflict as the first step in combating the drug problem, and it was to stay in effect until 2005. By 2001, Plan Colombia was not yielding the results that Clinton had hoped for. Human rights organizations, drug policy researchers, journalists, academics, and activists all observed coca cultivation increasing, the price of coca rising, and the street price of coca dropping. The failed Plan Colombia program and the unyielding problems beset by the democratization of the drug trade set the stage for a new counterinsurgency policy, one that was introduced by a hard-line presidential candidate, Alvaro Uribe.

During the 2001 campaign Uribe promised to execute a hard-line approach to insurgency, and his presidential victory served as a mandate to do just that. He promised to adhere to a law-and-order rhetoric as expressed in Colombia's *Democratic Defense and Security Policy* [1]. The attacks of September 11 in the United States drew greater attention to national security. It wasn't long before U.S. Attorney General John Ashcroft supported the position that *drug trafficking and terrorism are the same* [1]. Uribe would now present the Colombian counterinsurgency, once waged as the War on Drugs, as part of the Global War on Terrorism.

Uribe focused his attention on military recruiting to enlarge the size of Colombia's armed forces. He devised to fund this troop increase by legislating a war tax to increase revenues. And, he chose to undo the criticisms that Colombia's government had abdicated counterinsurgency to the paramilitaries by implementing a state-based approach to counterinsurgency.

All of these new efforts were introduced when Uribe assumed office in 2002. Logically, there would be a time lag before he would see the fruits of his efforts. The recruitment of soldiers would understandably take many months, given the fact that Uribe sought to increase the military from 125,000 to 225,000; however, by 2004 his military exceeded his original goal by 125,000 [3]. The collection of the 1.2 percent war tax would take place over the next few tax cycles, except that Uribe coupled the tax with a declaration of emergency powers, and that action yielded over $800 million with 70 percent on defense spending for 2003 [4]. Undoing the stigma attached to abdicating the counterinsurgency would take much for Uribe to negate. He was supported in this effort in 2002 when the United States authorized Uribe to use $1.7 billion to change the military balance and contain violence [1]. Partnering with the United States in the war against terrorism expanded the role, mission, and authority of the Colombian military. (The United States also encouraged a strategy of overwhelming military power and the use of the military for a quicker, more effective result versus a diplomatic approach.) Uribe would proceed with unilateral authority. At the same time the media projected derogatory images of Bin Laden and the Taliban as *Bin Ladenes* and *Talibanes* [1]. The Colombian population was now expected to view the fight against narco-terrorism as part of the Global War on Terrorism. Uribe was faced with an uphill battle, yet his government started to realize the results of its political and strategic changes in a relatively short period.

There is much in the way of quantifiable data that substantiate the effects of the policy change in Colombia. According to the U.S. State Department, between 2002 and 2006 Colombia saw decreases in homicides by 37 percent, kidnappings by 78 percent, terrorist attacks by 63 percent, and attacks on the country's infrastructure by 60 percent [5]. Uribe's approval rating rose with these successes.[5] Colombia's statistics show that in 2004 over 10,000 insurgents were either captured or killed and 3000 were disarmed [6]. This left the insurgents in a disjointed state, with 15,000 to 20,000 fighters on 105 fronts affecting 40 percent of the country [1].

Some Colombians may have mixed feelings of support for Uribe's administration, but a 2005 survey indicated that Latin Americans prefer strong, authoritarian governments. Coupled with that preference are the measurable successes of Uribe's counterinsurgency strategy, specifically the decreases in the number of insurgents and nefarious acts, all of which yielded 62 percent of the popular vote in favor of Uribe in the 2006 election [7].

MAPPING QUALITATIVE TO QUANTITATIVE

The description above provides a narrative on the evolution of Colombia's counterinsurgency. It completes the first task of the case study, which was to research the social sciences to assess Colombia's political activities and counterinsurgency strategy.

For the model to show the effects of policy change numerically, the analyst must somehow characterize each factor derived from the qualitative research in a quantifiable manner. It would be very simple to select only numerical data, such as the number of violent acts, media reports, or deaths. Additional numerically measurable variables speak to economic up- and downturns, money flow (laundering, drug trade), the number and types of tactics (the Dirty War), up- and downturns of insurgent numbers, government policies resulting in increases in revenue, refugee movement, and trade union membership. All of this is useful, quantifiable data; however, the analyst must be wary when doing a country study because these numbers are often unreliable. Moreover, relying on easily retrieved quantifiable data such as the number of internally displaced people, money

---

[5]There are different figures reflecting Uribe's approval rating in 2004: 80 percent (Crandal); 68 percent (Tiecher).

flow, and even the numbers of insurgents and peasants killed does not thoroughly express the gravity of the situation or portray the full multi-layered view of the situation at hand; and in the case of Colombia, the numerical data do not capture the underlying nature of the insurgency and the counterinsurgency efforts. *Simply, the numerical representations are devoid of context and interpretation*. That is why easily retrievable quantifiable data should be paired with qualitative data and analyzed jointly. So how does one do this? One suggested method, introduced in Chapter 2, is **mapping**.

Recall that the process of mapping is done by translating qualitative data into numerical values and mapping that assessment to a numerical representation or index value. This qualitative-to-quantitative mapping requires placing or binning the 128 factors (gleaned from the research on Colombia's insurgency) into categories more easily represented by an index value. An index value is chosen because it implies a reference value against which future changes may be compared. Thus, five indices were created that facilitated the incorporation of these factors into the overarching model of the Colombian case study. These indices also quantitatively represent the nature of the insurgency inside the country and the government's response to the insurgency. The five indices selected for this case study are (1) a polity index, (2) a human rights index, (3) a social capacity index, (4) a counterinsurgency index, and (5) an insurgency index. Indices 1 to 4 represent an assessment of government policy on the insurgency, while index 5 characterizes the influence of the insurgents on the overall system. The indices were defined as described below.

A **polity index** is a measure of the level of democracy of a country. It includes such things as the form of government (monarchy or democracy), a constitution and rule of law, a balance of power within the government, a system of checks and balances among branches or offices of the government, free elections, two-party or multiparty political representation, sound and independent judicial system, public support for the party in power and a venue for impeachment, civilian leadership of the military, and free and objective media. The polity index also includes cultural policies and conditions that refer to how a nation or state supports and accepts the cultural traditions of its population from both a majority and a minority standpoint. Culture includes such aspects as social behavior, beliefs, and traditions. Often, cultural intolerance by a government, or the perception of intolerance propagated by insurgents, is a factor in causing some members of a population to form or join an insurgent group against that government.

Economic policies and conditions are also part of the polity index, as they provide a measure of how well a nation or state is meeting the basic needs of its population. Such needs as food, clothing, shelter, medicine, and education are all necessary for a population to survive, let alone thrive. If a government fails to provide these basic needs, members of a population may form an insurgency in an attempt to force the government's hand in providing the basic economic necessities.

Another significant measure of the polity index is security policies and conditions, which go directly to a population's feeling of security. This is a significant factor influencing the formation of an insurgent effort. If a government is unable to provide an adequate level of security, the populace may try to take security responsibility into its own hands to provide the necessary level of protection, or the populace may feel coerced into supporting an organized insurgency that offers protection at the price of supporting the insurgency. Finally, the populace may remain passive in support of the government. Insurgents engage terrorist tactics to show the civilian population that the government cannot provide security. Without security, the civilian population is less likely to aid the government for fear of reprisal from insurgents.

A **human rights index** is a measure of the abuses of civil liberties and the degree of political marginalization experienced by a country's civilian population.

A **social capacity index** is a measure of the ability of the population to effect social change. The index also addresses religious policies and conditions, especially in the form of religious intolerance or oppression. This is a proximate cause for many insurgencies. For the civilian population the defense of religion against a perceived assault can legitimize an insurgency and sustain its morale. The civilian population can be outraged when the government is indifferent and/or adverse to their religious sensibilities.

A **counterinsurgency index** measures the will, capability, and commitment of a counterinsurgent effort. Specific to Colombia were abuses of ethnic violence and internally displaced people/refugees, unprotected porous national borders, targeted populations (politicians, labor, media, peasants, regional and urban dwellers), and vigilante justice that the counterinsurgency failed to address.

An **insurgency index** measures the strength of an insurgency and the factors that affect the insurgency. This index carefully weighs the intent of the insurgency and the number and structure (militaristic or egalitarian) of the combatants. The offensive approach, strategic or tactical, also includes an evaluation of the types of violent acts and against whom those acts are being targeted. This speaks to the nature (sadistic or Mafia-like) of the insurgency. It also addresses relationships between the insurgents and the civilian population and the insurgents and the government. The level of support domestically (recruitment) and internationally (financial) allows for speculation on the potential longevity of the insurgency. With the indices defined and the factors binned, a second division of factors occurred within the indices.

The factors under each index were divided into those that have a supportive influence or effect on the index and those that are detractors: that have a negative influence or negative effect. This facilitates the ability to compute each index by assigning a value to its respective factors according to their assessed significance. These values are then summed to provide a composite score. The factor values range from 1 to 5 (not significant to very significant). These assessed values are determined by a subject-matter expert's review of the applicable literature characterizing the nature of each factor. Because some factors represent a supportive influence and some represent a negative influence, the composite index score could conceivably sum to zero or have a negative value. However, what is important is that a composite numerical value can be computed to represent a qualitative assessment of those factors.

In this case study a Likert scale method provided the qualitative-to-quantitative scoring, the 1-to-5 value range, which captured the context of the insurgency and the conditions influencing it in a manner not represented by a pure count of some occurrences of violence or value of monetary flow. This index value represents an assessment for a specific period in time. It can be used to compare the state of an insurgency over time as the factors are rescored when changes occur. Thus, it provides a means for predictive analysis of policy changes on the insurgent effort.

Table 6-1 lists the supporting and detracting factors of the polity index between the years 1993 and 2001, when the counterinsurgency was waged as a War on Drugs, and then between the years 2001 and 2006, when the War on Drugs transitioned into the Global War on Terrorism. The table lists the polity index composite score between the years 1993 and 2001 as a negative because detractions to the measure of democracy result in a relatively ineffectual government. (This same tabling of data must be done with all five indices.) Table values of zero from 1993 to 2001 represent factors that were not assessed for that period of time, thus do not contribute to the scoring.

With the introduction of Uribe's new domestic and foreign policies, the polity index score changes post-2001. During the 2001–2006 period, Uribe's hard-line policy was well in place and being executed successfully; he was supported by the U.S. president with excellent funding (adding to his ability to recruit, train, and equip a larger counterinsurgency military), and there is a united effort between the two presidents to wage

TABLE 6-1   Polity Index Table

	1993–2001	2001–2006
Supporting factors		
A democracy (third oldest with popular, free elections, no long-term military rule, no tyranny)	5	5
Legitimacy as political institution with a constitution, 42 political parties	5	5
Legitimacy as perceived by people who support Uribe and civilian participation	4	4
Government policy of societal approach to counterinsurgency and Uribe's law-and-order rhetoric	0	4
Expanded representation in government	4	4
Grouping insurgency with terrorism empowers government to take a stronger approach	4	4
In 2002 there is a new Colombian president; a hard-line policy is now well in place and being executed	0	5
Uribe is supported by the U.S. president with excellent funding	0	5
There is a united effort between the two leaders to wage a counterinsurgency to regain Colombia	0	5
Plan Colombia is replaced by the Andean Counter Drug Initiative	0	5
The United States authorizes Uribe to use funds from the War on Drugs on the counterinsurgency	0	5
U.S. policy supports an expanding role for the military, emboldens the military	0	4
The United States supports the unilateral authority of Uribe	0	4
The United States promises ongoing support	0	5
The war tax that is implemented results in big dollars to the Colombian coffers	0	5
Post-9/11 there is a media/propaganda push that supports the government and labels drug traffickers as Bin Ladenes and Talibanes; the civilian population (those politically active) have accepted the cultural–societal shift from a drug war to a war against terrorism	0	5
Policies (land reform, economy) developed are executed, not with great success, but an effort made and shown over time	0	2
Uribe elected with 62% vote, although not an overwhelming majority as in 2002, he is able to continue his policies	0	4
Supporting factors total	**22**	**80**

TABLE 6-1 *(Continued)*

	1993–2001	2001–2006
Detractions		
Colombia cannot mature as a democracy with unending conflict	−5	−5
Legislature fragmented, ruled by elites who sponsor clientelism, relationship with the president tenuous	−5	−5
Judicial capacity	−5	−5
Government cannot provide for national domestic security; absence of state, no rule of law	−5	−3
In 1980–2004, over 100,000 people killed due to political conflict	−5	−5
Politicians corrupt (transparency international-corruption perception index 7.5/10) results in diversions of $1.76 billion from state budgets	−4	−4
Political history: two parties, people polarized with no sense of well-being	−4	−4
Checks and balances in government (power in the hands of a few)	−3	−3
Grouping Insurgency with terrorism compromises democracy	−4	−4
Stasis—accepting this as how things will be—the *politics of anesthesia*	−4	−4
Legitimacy as perceived by people who say that government has abdicated counterinsurgency	−3	−2
Institutions (government services—lack of institutions creates legitimacy for guerrillas)	−3	−2
Obstacle to advancing minorities in government is violence against their leaders	−3	−3
Spoilers (don't want peace, their power lies in the situation as is)	−2	−2
Detractions total	**−55**	**−49**
Index total	**−33**	**31**

a counterinsurgency to gain Colombia back from the insurgents. Additionally, Plan Colombia was replaced with a new funding program called the **Andean Counter Drug Initiative**, which ensured U.S. support of the Colombian counterinsurgency effort. Table 6-2 shows the newly computed polity index values for the period 2001–2006 in relation to 1993–2001. The new composite index shows an 85 percent positive change in policy factors affecting government resiliency from the polity standpoint. (This same tabling of data must be done with all five indices to reflect the change summary.)

TABLE 6-2   Polity Index Change Summary

Factor Summary	Factor Score, 1993–2001	Factor Score, 2001–2006
Supporting factors	+22	+80
Detractions	−55	−49
Composite Score	−33	+31

When all five indices have been expressed in tables, the qualitative-to-quantitative assessment of Colombia's insurgency is complete. Next is the creation of a model using system dynamics as the modeling tool. Borrowing from an insurgency model developed previously by Sokolowski and Banks, the Colombian data can be expressed in terms similar to those depicted in this model, but with certain modifications to incorporate the newly developed indices (see Figure 6-1) [8]. This system dynamics insurgency model depicts the main variables that influence insurgency behavior. The model can be divided into two sets of variables. The first set of variables, which describes government policies, represents the propensity of the government to collapse into an insurgency situation. The propensity is dominated by a combination of cultural, religious, economic, and security conditions and policies within the state. The second set of variables represents the insurgents' efforts to recruit new members to their cause. These variables include violent acts, propaganda, and social network influence.

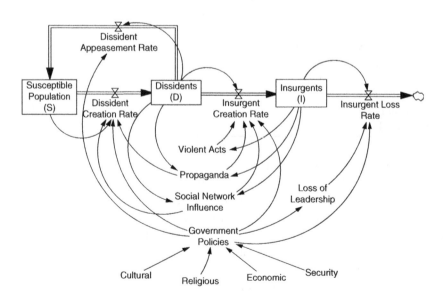

Figure 6-1   System dynamics insurgency model.

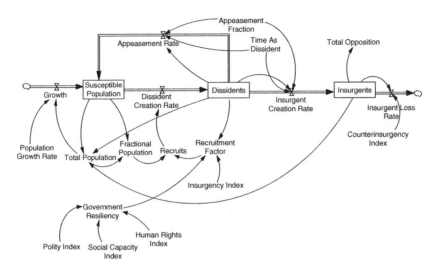

Figure 6-2 Revised system dynamics insurgency model.

This model became the starting point for analyzing the Colombian case study but was not sufficient to capture the quantitative inputs needed to show the changes in the three populations (susceptible, dissident, insurgent) over time. The indices, on the other hand, allow for this quantitative representation. Thus, the model of Figure 6-1 was updated to incorporate those indices. The factors in Table 6-1 can now be matched to the propensity factors noted above, and those for the other index tables can be matched to the government and insurgent conditions of Figure 6-1. What resulted was a revised system dynamics insurgency model (Figure 6-2). This model shows the three populations of the original model: susceptible population, dissidents, and insurgents. The creation and loss rates are now governed by the index values. Government resiliency directly affects the propensity for dissident recruitment and competes with the effectiveness of the dissidents and insurgents on recruiting followers determined from the insurgency index. The counterinsurgency index indicated the government's effectiveness in removing insurgent members.

SYSTEM DYNAMICS

Chapter 3 comprised a lengthy discussion of system dynamics as a modeling method. That presentation also detailed the process by which one develops a system dynamics model and the data required to fully engage the model. Creation of the research methodology should include thoughts about articulating the research findings, modeling the data, proffering suggestions, and stating solutions to the research question. Thus far, two steps in this case study methodology have been completed: the qualitative discovery and the conversion of qualitative to quantitative values. The third step is incorporating the data into the model of Figure 6-2 to have a specific representation of the Colombian insurgency for a given period of time. As noted, there are two distinct periods of insurgency activity identified in this study: 1993–2001 and 2001–2006. For these time periods actual insurgent strength data were identified. Figure 6-3 provides a graphical representation of those data.

The first time period (1993–2001) was used to calibrate the system dynamics model of Figure 6-2. The calibration was performed by fitting the equations that define the system dynamics model to actual insurgent strength. The results of the curve fitting are shown in Figure 6-4, where the solid line represents the output of the mathematical model embedded in the system dynamics representation. Now that the model has been calibrated, a reference set of indices exist and can be related to their respective composite scores. Table 6-3 shows the reference set of indices computed from the calibration and the associated composite scores.

As the government implements new policies and the insurgents try new strategies, the index factors will change and must be reevaluated. That was the case for the second time period, 2001–2006. The composite scores for each index were recalculated as shown in Table 6-2. These new scores can be used to determine new index values for the system dynamics model. This essentially changes its mathematical characteristics to show what the expected effect on insurgent level should be. How does

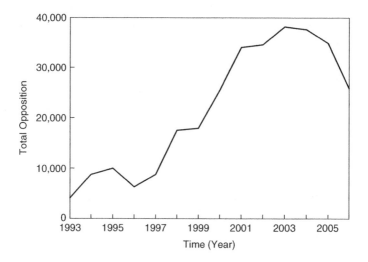

Figure 6-3   Actual insurgent strength.

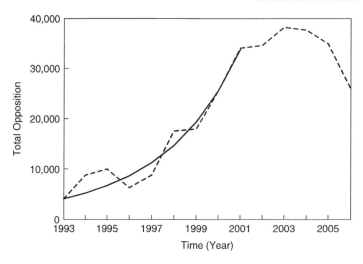

Figure 6-4   Insurgency model calibration.

TABLE 6-3   Insurgency Model Calibrated Index Values

Index, 1993–2001	Index Value	Index Composite Score
Polity index	1.0600	−33
Social capacity index	1.0594	−45
Human rights index	1.0205	−58
Counterinsurgency index	−0.02915	+10.5
Insurgency index	1.9343	+61

TABLE 6-4   Index Value Summary

Index	Index Composite Score		New Index Value
	1993–2001	2001–2006	
Polity index	−33	+31	1.9557
Social capacity index	−45	−45	1.0594
Human rights index	−58	−44	1.3573
Counterinsurgency index	+10.5	+24.5	0.0387
Insurgency index	+61	+42	1.3308

the model of Figure 6-2 allow for the quantitative prediction of future insurgent strength? Table 6-4 shows the recomputed index values, and Figure 6-5 shows how the model responds and predicts the effect on the insurgency. The solid line in the figure is the model's predicted response to overall changes in the system.

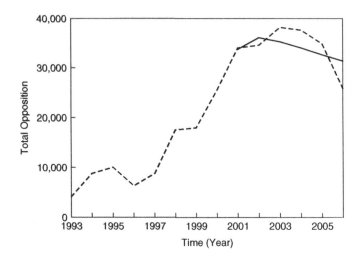

Figure 6-5   Insurgency model prediction curve.

RESPONDING TO THE RESEARCH QUESTION

The research question for this case study asked how to measure and represent governmental policy changes in Colombia's War on Drugs in a model that can predict insurgency strength and prescribe counterinsurgency strategy. The answer came as a formal modeling methodology used to analyze the strength of an insurgency within a state by qualitatively assessing numerous factors that contributed to influencing insurgent behavior. This method also provided for mapping the qualitative assessment into a quantitative representation without losing the context of the situation. The qualitative scoring was used in defining five indices that characterized the insurgency. The indices were included in a system dynamics model to show changes in insurgent strength in response to changing government policy and insurgent capacity.

The study revised a previously developed insurgency system dynamics model to better capture the factors influencing the insurgency in a quantitative manner, and it defined a set of factors used to evaluate the Colombian insurgency in a consistent manner. The methodology included:

1. Conducting a historical review of the insurgency to identify the contributing factors
2. Performing a qualitative-to-quantitative mapping of these factors using a qualitative assessment strategy such as the Likert scale rating

3. Using the qualitative representation to compute index values for the model
4. Calibrating the model to known data for a specific time period
5. Reassessing factors based on actual or contemplated policy changes and recomputed the index values
6. Rerunning the model with the new indices to analyze the change in insurgency strength based on the revised information

Why is system dynamics the appropriate modeling tool for this case study? *Because a system dynamics model can depict the main variables that influence insurgency behavior.* It accommodates a contextual, interpretative analysis of the qualitative research. From this study one can see that this methodology could be used to analyze contemplated policy changes and their temporal affect on insurgency strength.

What are the strengths and weaknesses of using system dynamics (specific to this research question)? Probably the most significant strength with this type of modeling is that the analysis of Colombia's insurgency can be perpetual because the index values used represented an assessment for a specific period in time. Therefore, these index values can be reassessed to compare the state of the insurgency with rescoring of the factors when changes occur. This type of modeling provides a means for prescriptive and predictive analysis of policy changes for the counterinsurgent effort.

What other case studies can be reviewed via a system dynamics approach? Essentially, this approach can be engaged in any domain where one may want to explore policy implementation. For example, system dynamics can be used for modeling various oil production policies and their impact on the global economy. The model would then provide a venue to test specific policies and possibly provide insight into unintended consequences of a particular policy that would not be obvious from the policy formulation itself.

Colombia's insurgency is especially dangerous because the incentive for the insurgency funds the insurgency, and that incentive—drug trafficking—is one that seemingly cannot be stopped. The democratization of the drug trade as a result of the dissolution of the two large cartels in the mid-1990s spawned many splinter groups. For decades the government's counterinsurgency effort could not match the resources of the insurgency. That changed in 2002, however, when Colombia's President Uribe began executing a state-based, hard-line approach to insurgency. Shifting the War on Drugs to the Global War on Terrorism facilitated significant changes. Not long after the implementation of Uribe's counterinsurgency policy, Colombia began to realize changes in the capacity of the insurgency. This model measured and represented the effects of those government policy changes.

KEY TERMS

Dirty War	Plan Colombia
democratization of the drug trade	mapping
Conservative Party	polity index
Liberal Party	human rights index
La Violencia	social capacity index
National Front	counterinsurgency index
FARC	insurgency index
ELN	Andean Counter Drug Initiative
AUC	

REFERENCES

[1] Murillo MA. *Colombia and the United States: War, Unrest, and Destabilization*. New York: Seven Stories Press, 2004.

[2] Burt JM, Mauceri P, eds. *Politics in the Andes: Identity, Conflict, Reform*. Pittsburgh, PA: University of Pittsburgh Press, 2004.

[3] Council on Foreign Relations: Report for the Independent Commission. *Andes 2020: A New Strategy for the Challenges of Colombia and the Region*. Washington, DC: US Government Printing Office–COFR, 2004.

[4] Teicher DE. *The Decisive Phase of Colombia's War on Narco-terrorism*. Maxwell AFB, AL: USAF Counter-proliferation Center, 2005.

[5] US State Department. http://www.state.gov/p/wha/ci/co/.

[6] Crandal R, et al., eds. *The Andes in Focus: Security, Democracy, and Economic Reform*. Boulder, CO: Lynne Rienner, 2005.

[7] Wiarda HJ. *Dilemmas of Democracy in Latin America: Crises and Opportunity*. Lanham, MD: Rowman & Littlefield, 2005.

[8] Sokolowski JA, Banks CM. From empirical data to mathematical model: using population dynamics to characterize insurgencies. In: *Proceedings of the 2007 Summer Simulation Multiconference*, San Diego, CA, July 15–18, 2007, pp. 1120–1127.

CASE STUDY BIBLIOGRAPHY

Alesina, Alberto, ed. *Institutional Reforms: The Case of Colombia*. Cambridge, MA: MIT Press, 2005.

Arboleda, Jairo, et al. *Voices of the Poor in Colombia: Strengthening Livelihoods, Families, and Communities*. New York: World Bank, 2004.

Bailey, John, and Lucia Dammert, eds. *Public Security and Reform in the Americas*. Pittsburgh, PA: University of Pittsburgh Press, 2006.

Burt, Jo Marie, and Philip Mauceri, eds. *Politics in the Andes: Identity, Conflict, Reform*. Pittsburgh, PA: University of Pittsburgh Press, 2004.

Cott, Donna L. *From Movements to Parties in Latin America: The Evolution of Ethnic Politics*. Cambridge, UK: Cambridge University Press, 2005.

Council on Foreign Relations: Report for the Independent Commission. *Andes 2020: A New Strategy for the Challenges of Colombia and the Region*. Washington, DC: U.S. Government Printing Office–COFR, 2004.

Cragin, Kim, and Bruce Hoffman. *Arms Trafficking and Colombia*. Santa Monica, CA: Rand Publications, 2003.

Crandal, Russel, et al., eds. *The Andes in Focus: Security, Democracy, and Economic Reform*. Boulder, CO: Lynne Rienner, 2005.

Crocker, Chester A., et al. *Grasping the Nettle: Analyzing Cases of Intractable Conflict*. Washington, DC: United States Institute of Peace Press, 2005.

Crotty, William, ed. *Democratic Development and Political Terrorism: The Global Perspective*. Boston: Northeastern University Press, 2005.

Dudley, Steven. *Walking Ghosts: Murder and Guerrilla Politics in Colombia*. New York: Routledge, 2004.

Gilderhus, Mark T. *The Second Century: US–Latin American Relations Since 1889*. Wilmington, DE: Scholarly Resources, 2000.

Lowenthal, Abraham F., and J. Samuel Fitch, eds. *Armies and Politics in Latin America*. New York: Holmes & Meier, 1986.

Mares, David R. *Violent Peace: Militarized Interstate Bargaining in Latin America*. New York: Columbia University Press, 2001.

McPherson, Alan. *Intimate Ties, Bitter Struggles: The United States and Latin America Since 1945*. Washington, DC: Potomac Books, 2006.

Murillo, Mario A. *Colombia and the United States: War, Unrest, and Destabilization*. New York: Seven Stories Press, 2004.

Newman, Edward, and O. Richmond, eds. *Challenges to Peace Building: Managing Spoilers During Conflict Resolution*. New York: United Nations University, 2006.

Palmer, David S. *US Relations with Latin America During the Clinton Years*. Gainesville, FL: University of Florida Press, 2006.

Rabasa, Angela, and Peter Chalk. *Colombian Labyrinth: The Synergy of Drugs and Insurgency and Its Implications for Regional Stability*. Santa Monica, CA: Rand Publications, 2001.

Rochlin, James F. *Vanguard Revolutionaries in Latin America: Peru, Colombia, Mexico*. Boulder, CO: Lynne Rienner, 2003.

Rojas, Cristina, and Judy Meltzer, eds. *Elusive Peace: International, National, and Local Dimensions of Conflict in Colombia*. New York: Palgrave Macmillan, 2005.

Shaw, Carolyn M. *Cooperation, Conflict, and Consensus in the Organization of American States*. New York: Palgrave Macmillan, 2004.

Smith, Hazel, and Paul Stares, eds. *Diasporas in Conflict: Peace-Makers or Peace-Wreckers?* New York: United Nations University Press, 2007.

Teicher, Dario E. *The Decisive Phase of Colombia's War on Narco-terrorism*. Maxwell AFB, AL: USAF Counter-proliferation Center, 2005.

United Nations. *Global Agenda: Issues Before the 58th General Assembly of the United Nations*. New York: United Nations Association, 2004.

Vincent, Marc, and B. R. Sorensen, eds. *Caught Between Borders: Response Strategies of the Internally Displaced*. London: Pluto Press, 2001.

Wiarda, Howard J. *Dilemmas of Democracy in Latin America: Crises and Opportunity*. Lanham, MD: Rowman & Littlefield, 2005.

7 Case Study: The Polish Solidarity Movement—Laying the Foundation for the Collapse of Soviet Communism

INTRODUCTION

The global environment that we share today stems from an international system that took root in 1945 and dissipated in 1989. This period of history known as the **Cold War** centered on two ideologies and two superpowers: on the one side was the United States and its democratic, capitalist ideals, on the other side was the Soviet Union and Communism. For many who study history, foreign policy, and collective security, the Cold War is regarded as an aberration—an orderly chaos comprised of containment, NATO, the Warsaw Pact, mutually assured destruction (MAD), domino theories, arms races, missile gaps, brinkmanship, deterrence, compellence, SALT (strategic arms limitations talks), START (strategic arms reductions talks), CTBT (comprehensive test ban treaty)—the list of catch phrases, policies, and treaties goes on. The effect of the collapse of Soviet Communism is profound, and it is ever present in the current geopolitical behavior of states throughout the world that participated directly or indirectly in the Soviet system.

There is a plethora of literature reflecting on this remarkable period in world history. One popular text written in 1997 summarized the perception, the reality, and the lessons learned from the Cold War: *We Now Know: Rethinking the Cold War History* by John Lewis Gaddis, professor of history at Yale University. Gaddis spent a year as a lecturer at the University of Oxford, where he taught the history of the Cold War from two

Modeling and Simulation for Analyzing Global Events, By John A. Sokolowski and Catherine M. Banks
Copyright © 2009 John Wiley & Sons, Inc.

perspectives. First, he had the advantage of teaching the war from start to finish, 1945–1989; and second, he could now teach it as international history, steering away from his usual bipolar Soviet–U.S. representation. This book, a result of his teaching term at Oxford, is a comprehensive history of this 50-year chronicle, which includes discussions on domestic and world economics, ideology, the role of Third World countries and nuclear proliferation, the Cuban Missile Crisis, and the issue of compellence. Gaddis remains current in his analysis of the Cold War. In 2005 he introduced new information drawn from recently opened Chinese, East European, and Soviet archives in his text *The Cold War: A New History*.

Another authority on the subject of the Cold War is Timothy Garton Ash, professor of European Studies at the University of Oxford. His research has chronicled the transformation of Europe over the last quarter-century. Ash refers to his political writing as *history of the present*. He has published two works that are applicable to this case study: *The Magic Lantern: The Revolution of 1989 Witnessed in Warsaw, Budapest, Berlin, and Prague* (published in 1993) and *The Polish Revolution: Solidarity, 1980–82* (published in 1984). The challenge for many who choose to research the subject of the Cold War lies in sorting through the literature and information available. There are numerous websites and educational and teaching materials, volumes of documents and primary sources, biographies and memoirs, histories, and documentaries—all providing a body of knowledge that continues to expand in size and in understanding of this epoch of world history.

Although there is no question that the Cold War waged between the United States and the Soviet Union shaped today's global environment, many people are still not clear regarding the specific events that shaped, affected, and caused the collapse of Soviet Communism. In this chapter we focus on personalities and incidents inside the republic of Poland and the founding of the Solidarity movement in 1980. We explain what happened, why things happened, and how the incidents were related. The analysis will take a multifaceted perspective to elucidate the complexity of diplomatic relationships, economics, the role of the military, the responsibility of the Catholic church, and the human capacity for freedom. The Solidarity movement was the basis for the eventual collapse of Poland's Communist regime and all that Communism entailed. This story is another in the human struggle for self-determination and sociocultural freedom. No person, movement, regime, or church caused the demise of Communism in Poland and in the greater eastern European states. It was the *totality* of the events in each Eastern Bloc country and the collective efforts of the Western nations (France, West Germany, Great Britain, and United States) that serve as a basis for Soviet Communism's demise. This case study looks at Poland as the leading Warsaw Pact state that set the stage for the cumulative events of 1989 through its own particular social revolution, Solidarity.[1]

The chapter is presented in four subsections: (1) a structured methodology to represent human behavior modeling of the Solidarity movement using agents, (2) a concise narrative of Solidarity as a social revolution that laid the foundation for the eventual collapse of Soviet Communism as expressed through primary and secondary documents, (3) an explanation of human behavior modeling and the use of agents—*who is the agent and what is the environment*—to model human behavior with agents, and (4) a response to the research question.

DEVELOPING THE RESEARCH QUESTION AND METHODOLOGY

The case study requires an analysis of the Polish Solidarity Movement and the social revolution that took root in 1980 and evolved during that decade. Poland's revolution instigated the collapse of the state's Communist regime with a combination of popular protest and nonviolent negotiations. The revolution was the culmination of an economic downward spiral that saw massive increases (as much as 200 percent) in the price of food and consumer goods, and strikes from laborers at factories, shipyards, and mines. The workers' strikes gave rise to a trade union that challenged the state's control of the politics, economy, and daily lives of the Polish people. Among the Warsaw Pact states it was Poland that took the lead in negotiating with the Communist authorities. By the late 1980s, each Warsaw Pact state had learned from the Polish experience. Coupled with Moscow's new philosophy of *openness*, each country took its respective turn at its own revolution, adding weight to the already crumbling Soviet system.

In this study we explore the structural and social components of the Solidarity movement. The structural components include the economy, the political system, and the military, which were all under the control of the Polish Communist regime. The social component includes the labor movement, led by Lech Walesa, the Polish people (shaped by their history, culture, and religion), and the Catholic church and Pope John Paul II. The goal of the research is to understand the relationships between and among

---

[1]Warsaw Pact member states included Soviet Union, Poland, East Germany, Czechoslovakia, Hungary, Romania, Bulgaria, and Albania.

the structural and social components individually and collectively. These relationships shaped behaviors and outcomes.

What one can expect to learn from this case study is that revolution doesn't happen in a vacuum: It is the coalescence of structural and social phenomena and the culmination of a **circular environment** in which action is based on current conditions ... which are affected by actions ... which change conditions ... . The research question for this case study takes this obvious lesson to a higher level: *Using the Polish Solidarity movement and revolution as a case study, how does modeling human behavior with agents capture (1) both the action and/or reaction of agents to stimuli from the environment, and (2) the relationships between and among the agents to provide a multifaceted examination of the social revolution that led to the collapse of Communism in Poland?*

Integral to this case study is a review of primary and secondary sources to describe the social structure as well as the economic, political, and military structure of Poland to understand the conditions that proved ripe for revolution. Also, included is a review of historical and biographical data to understand the individual actors, the culture, and the role of religion. The research will then engage M&S as a tool in developing a representative human behavior model (termed *modeling human behavior with agents*) that includes the empirical data to substantiate the model. The goal is to develop a human behavior–agent model that characterizes and communicates the complexity of revolution. With the research question in hand, the modeler can employ the model development methodology outlined in Chapter 4 to construct an agent-based model.

The discussion below is based on qualitative research from primary and secondary sources on the subject of the Solidarity movement and the social revolution in Poland. A bibliography for this case study appears at the end of the chapter.

BACKGROUND: QUALITATIVE RESEARCH

The collapse of Soviet Communism in Eastern Europe was the result of simultaneous revolutions in eight states that advanced with great speed once the first domino fell. Poland, Hungary, East Germany, Czechoslovakia, Romania, Yugoslavia, Albania, and Bulgaria all experienced, in part or in whole, what has variously been called a system of socialism, totalitarianism, Stalinism, polit-bureaucratic dictatorship, real existing socialism, state capitalism, and dictatorship; yet, within the course of one year these states would realize a dramatic change in their respective political systems. Dubbed the *Year of Revolutions, European Revolutions*, and the

Spring of Nations, 1989 developed into a year of solidarity within and among nations [1]. During the Velvet Revolution in Czechoslovakia, Timothy Garton Ash emphasized the speed with which things moved when he quipped to organizer Vaclav Havel: "[The Revolution] took ten years in Poland, ten months in Hungary, ten weeks in East Germany, and ten days in Czechoslovakia" [1].[2] This regional sense of solidarity existed because all the states consisted of the same structural and social components.

The social components differed from state to state but always included three prominent variables: the Catholic church, the people (history and culture), and labor (workforce). The structural components, such as the economy, political system, and the military, were controlled by the regimes. What made Poland unique was that its three prominent social components were so closely tied that they could not be separated. The Polish people were in solidarity with their history, their Slavic roots and language, and the 1000-year-old Polish Catholic church.[3] How did this Polish *oneness* expand into a solidarity that represented a labor revolution? What caused the revolt, and who led it? Importantly, how did the Communist leadership respond? Solidarity would evolve into a moral and social movement deeply rooted in the workforce; it was a response to the structural stimuli, the economy, the **apparat** (a vast bureaucratic web of power that housed the organs of the ruling Communist party), and the military (both Soviet and Polish)—all of which offered nothing but despair to its constituents. The participants or actors included lay-left politicians, union leaders, laborers, students, intelligentsia, and church leaders. This conglomeration of Polish society responded to the structural stimuli in ways that were unprecedented and heroic.

The Polish State: Social

The Church   Polish history stretches back to the close of first millennium C.E., when the Polanie Tribe, the western branch of the Slavic peoples, migrated from central Eurasia and settled along the Vistula (Wisla) River. Constant pressure from the German tribes and Christian missionaries convinced the ruler of Polanie, Mieczyslaw I, to accept Christianity for himself and his people. By 1000 C.E. an independent Polish church organization was set up with the agreement of Otto III of Germany. The Polish church resembled the Czech system more closely than the German system;

---

[2]The revolution in Czechoslovakia actually took 24 days. It proceeded so quickly because it was the last revolution, and the Czechs had the benefit of the momentum set in place by the other states.

[3]In 966 C.E., King Mieczslaw was baptized, making Poland the bulwark of the Latin church.

thus, it could turn directly to Rome and the pope for protection without German interference. In 1024 C.E., Mieczyslaw's son, Boleslaw Chrobry (the Brave), became the first king of Poland, establishing Poland's right as an independent kingdom. From these national origins one can appreciate two significant facts regarding the Polish nation: (1) the long-standing presence of the Polish people among the Europeans, and (2) the close, enduring tie between Poland and the Latin church. By the fifteenth century, a golden age of democracy paved the way for a very tolerant Poland: It became a haven for heretics and religious diversity, and its patriotism and closeness to the church deepened. In the succeeding centuries the Catholic masses prayed to the Mother of God as the Queen of Poland, while the intelligentsia kept alive Polish culture and values—instilling a sense of *Polishness*. The nation of Poland and the Latin Catholic church were inseparable—this is key to understanding the culture of Polish society in terms of the Solidarity movement, whose seeds were sown on the heels of World War II.

Stalin's repression of Polish society made its way into the Catholic church. From 1949 through 1955 the Soviet-backed Polish government tried to subjugate the church, but failed. The church offered Poles an alternative ideology and worldview to that of Communism. The Polish Catholic church, led by Cardinal Stefan Wyszynski, fought atheism. The cardinal had an outstanding personality and was considered a skillful leader. His straightforward condemnation of Soviet repression earned for him a stay in prison as a dissident. He charged that it was madness to demand that "a whole nation renounce Christianity only because a small group of people believe the reconstruction of society is impossible without a materialistic ideology" [2]. Yet whenever the apparat experienced uprisings, it looked to the church to assist in quashing them; this, in effect, empowered the church. Throughout the 1960s (1959–1966), President Gomulka, jealous of the church's ability to attract and affect parishioners, declared war on Wyszynski and his cadre [3]. This war backfired and, instead, served to unite the people. Although there was never an open confrontation, the church was cautious in acting as a protective umbrella for Catholic intellectuals. It was during this time that Cardinal Wyszynski introduced the notion of Polish solidarity; he would unite all classes while defending the people's rights and Poland's rights. They, in turn, would be in solidarity: as Poles, as Catholics, and as human beings [2].

Two vital bastions existed in Poland during the 1970s: private agriculture and the Church. In 1968, Poland's two Cardinals, Karol Wojtyla (in Krakow) and Stefan Wyszynski (in Warsaw), listed the God-given human and civil rights that the government must respect. This message came from the pulpit, which meant that the vast majority of Poles, regardless of class, heard this message and now had a common vocabulary. Wojtyla's relationship with academics earned him a firm place with the intelligentsia. His role in both worlds (as pastor and professor) resulted in the alignment of the church and non-Communist intellectuals who strove to keep alive Poland's national conscience.

In October 1978, Karol Cardinal Wojtyla became Pope John Paul II: a Polish pope. The people wept with joy; the government groaned with despair. The Polish Communist party and apparat knew this pope very well. He was close to Cardinal Wyszynski, who they had dubbed the "Fox of Europe." Wyszynski had outwitted Russian commissars, Nazis, and Stalinists for over forty years; and it was Wyszynski who had groomed then-Cardinal Wojtyla. The Polish government viewed Wojtyla as the "thorny Archbishop of Krakow" [2]. Just prior to his election as pope, Wojtyla had denounced state censorship. Thus, a Polish sense of destiny began to surface, and there was a real and perceived confirmation of the church's authority with the election of Wojtyla [4]. The new pope vowed to make the dignity of labor an issue high on his agenda. John Paul's role in the events in Poland exerted extraordinary influence. He would offer Poland and the world an alternative to political and social oppression—he introduced a principle that spoke to the dignity of the person [5].[4]

The Politburo convened as soon as it learned of Wojtyla's election as pope to determine the position the apparat would take toward him, but their course of action was none other than to congratulate Wojtyla and recognize him as a son of Poland. Within weeks of his investment, John Paul II discussed an official Vatican trip to Poland to commemorate the 900-year anniversary of the martyrdom of Poland's St. Stanislaw, the first Bishop of Warsaw. The regime saw this as a nightmare because St. Stanislaw was a dissident! Despite much opposition from the Polish government, John Paul II negotiated successfully for a nine-day trip in June 1979.

The pope's visit to Poland in 1979 set the tone for this pope's role in the geopolitical arena. Having survived the Nazi invasion and Soviet domination, John Paul II was prepared as a member of the Catholic hierarchy in Stalinist Poland to take a central position. Here was a Slavic pope turning his attention to *mittel Europa*, the central bloc of Europe's nations and the underbelly of the Soviet Union [2]. He openly endorsed a fruitful synthesis of love of country and love of church [6]. The pope spoke of the special lesson that Christian Poland had to teach the world

---

[4]John Paul II believed that the tragedy of atheist humanism was that it stripped human beings of their transcendental character, destroying their ultimate significance.

and the responsibility that was placed upon this younger generation [6]. He taught the Poles that they needed to display a *maturity* that would move them forward as nonconformists [6]. As a nation, the Poles endured five years of Nazi rule and thirty-three years of Communism—in 1979, Pope John Paul II reclaimed Poland's history [7]. His visit was at a time when the economic crisis was deepening, yet he began a national regeneration [8]. The pope defined common values and unified Polish society to see beyond its potential and to respond to actual conditions [6]. On a much larger scale, he challenged the status quo of world order [2].

In 1979 an underground paper devoted an entire issue to a *Charter of Worker's Rights*. With that came a tacit alliance among the workers, intelligentsia, and the church [6]. The former Archbishop of Krakow, Karol Wojtyla, had served his parishioners for thirty years, teaching them self-regulation and self-limitation. With the establishment of the Solidarity movement, that alliance had an opportunity to test whether any revolution in the modern world could be both self-regulating and nonviolent [8].

The People   Revolution is something the Poles know quite well. Having been divided three times in the eighteenth century (1771, 1793, and 1795), the Poles conducted insurrections against the Russian czars in 1794, 1830, 1863, and 1905. They fought their oppressors, the Russian Orthodox church, and the German Protestants with religious zeal, but failed to maintain their autonomy. At the close of World War I (1918), Poland was established as a republic. It immediately had to defend its right to that status in 1920 during the Polish–Soviet War (from 1921 to 1939, Poland experienced its second republic). In 1939 the Soviets signed a Non-Aggression Pact with the Nazis that facilitated a one-front war for the German invasion of Poland's western front in 1939 (followed by the Soviet invasion in the east). Fierce opposition and resistance to German occupation did not earn for Poland its return as a republic.[5] Instead, with the close of World War II came Soviet occupation and domination. The **Sovietization** of Poland began in 1946 with the undeclared declaration of the Cold War [6]. The Poles reflect sadly on this war—it is a war they lost twice.

---

[5]The August 1–October 2, 1944 uprising in Warsaw was a heroic and tragic fight to liberate the city from Nazi occupation, undertaken by the Home Army at the time that Allied troops were breaking through the Normandy defenses and the Red Army had reached the Vistula River. Warsaw could have been one of the first European capitals liberated; however, various military and political miscalculations and geopolitics on the part of Stalin, Churchill, and Roosevelt precluded it.

The Poles suffered tremendous losses: over one-half million fighting men and women; six million civilians (22 percent of the population) of whom 50 percent were Christian and 50 percent were Jewish; nearly six million of the total Polish war losses were victims of prisons, death camps, raids, executions, annihilation in ghettos, epidemics, starvation, and excessive work. The country lost 38 percent of its national assets, with much of the eastern region swallowed up by the Soviet Union. Stalin's plan for his newly acquired Soviets was to institute the Communist system of governance and culture. He recognized early-on the challenge in accomplishing that with Poland, due to its fundamental and historic opposition and incompatibility with Soviet Communism. Thus, for thirty-three years Poland was led by a Communist party that struggled to maintain its strength. This history of resistance against the Soviets depicts the Polish national identity as a European people with a thirst for freedom and independence; and it is a long history of opposites: individualism versus collectivism; Catholicism versus Orthodoxy; democracy versus despotism [6].

The United States was opposed to the Communist regime in Poland but recognized it as one of the weakest in the Soviet sphere. As a consequence, the party's weakness made for a less oppressed Poland. The economic catastrophe during the 1970s and the civil disturbances throughout the country caught the attention of the Carter Administration (1977–1981). There was also a period of **détente** (a relaxation in international affairs), which allowed for discussions between West and East, with both groups stating their goals, such as recognizing two German states, American reduction of forces in Europe, and human rights. This era of détente had the effect of chipping away at Communist policy because it suggested an openness between the states that exposed the failures of the state parties, in particular the Gomulka government in Poland [9].

Poland displayed an internal liberalism that appealed to President Carter, and it served as a real-time case study on human rights. In fact, Poland was selected as the first presidential visit for Carter, no doubt highly recommended by his Polish-American National Security Advisor Zbigniew Brzezinski and second-generation Polish-American Secretary of State, Edmund Muskie. Carter was advised that the Soviets would view this as a provocation [10]. Brzezinski accompanied the President, arriving in Warsaw on December 29, 1977. Brzezinski made a visit to the Polish primate Wyszynski and offered him a note from President Carter that shared his best wishes, prayers, and faith. Carter also made known to the Cardinal that he admired what the Polish church represented; he was seeking the same goal [10]. Carter fully supported Solidarity and made a point of stressing U.S. political support for Polish independence. The United States supported Poland financially by providing commodity credits and funding: $400 million in 1979, $500 million in 1980, and $670 million in 1981 [10].

Throughout 1980, Polish workers came together to form a massive labor union called **Solidarity** (further discussion below). Brzezinski kept a close watch on this union and the events in Poland; he recognized diminishing Soviet control and a weakened Polish Communist party. The seriousness of the strikes and the internal commotion throughout the country led Brzezinski to believe that the Soviets would not allow these problems to persist—perhaps an invasion like that of Czechoslovakia 1968 was on their minds.[6] Brzezinski urged Carter to issue a strong statement condemning any Soviet invasion as well as allowing the Poles to address their problems independently. Carter brought the European allies into the discussion and all expressed their fear of the far-reaching consequences were the Soviets to invade. The United States closely monitored every movement of Soviet troops along Poland's borders, alerting the pope and the leadership of Solidarity. Brzezinski had no reservations about encouraging Polish resistance since they would not be taken by surprise [10].

The Reagan Administration (1981–1988) continued the Poland policy set out by Carter. President Reagan considered Solidarity a model for reform activities and supported it. However, with the declaration of martial law in December 1981, Reagan was compelled to change strategies as a way to sanction the Polish government for this action.

Labor (Workforce) With their history of fighting for national independence, it should come as no surprise that the Poles would resist exploitation of labor and the freedoms denied workers within the Communist system. Although described among the Soviet states as a workers' paradise, Polish laborers had another story to tell and did so via riots, demonstrations, and strikes [11]. This no doubt shocked many in Poland's sister states. The first demonstrations erupted in 1956 when workers discovered that they had been cheated of their wages. Thousands demonstrated in Poznan. When they were attacked, they rioted, calling for *bread and freedom*. The **Poznan Riots** brought about the articulation of two freedoms that would remain part of the workers' mantra: (1) worker's freedom from want, and (2) the nation's freedom from foreign domination. At the Kremlin, Nikita Khrushchev was reassured by Polish President Gomulka that he could control the situation, thus avoiding a Soviet invasion such as the one in Hungary. His ability to restore order and his return to power

---

[6]In August 21, 1968 the Soviet army invaded Czechoslovakia, along with troops from Bulgaria, East Germany, Poland, and Hungary. The occupation was the beginning of the end for the Czechoslovak reform movement known as the *Prague Spring*.

afforded him a new, more independent relationship with the Soviet leadership. For the Poles, this **Polish October** led to the end of Stalinism in Poland. The Polish October brought with it cultural freedoms, the release of political prisoners, the release of Warsaw Cardinal Stefan Wyszynski (which brought about renewed normalization of relations between the state and the church). Unfortunately, these freedoms did not last long. A gradual shift in the opposite direction saw a tightening of restrictions in 1959. The intelligentsia led protests opposing the government crackdown in 1964 and 1968, both of which were forcibly put down. Workers in Gdansk, Gdynia, and Szczecin (Baltic cities) responded to announcements of price increases for food with loud protests in the 1970s; these demonstrations were violently repressed. For the Poles this was anathema—Pole shall not kill Pole! Those who lost their lives were viewed as martyrs. The riots were quelled by Edward Gierek, who then replaced party leader Gomulka. Gierek promised to prevent the price increases, discuss reforms, and introduce a policy of repaid industrialization based on Western imports and credits. This resulted in an artificial rise of living standards and an economic policy that would eventually bankrupt Poland. For the workers, the 1970 riots paved the way for the working class to flex their muscles, place demands on the party (to include halting price increases, recognition of strike committees, representation in the party, independent trade unions), and channel their protests through the party leader.

The events in 1970 also set the tone for the much larger demonstration of workers' rights and demands. In 1980, Pope John Paul II wrote an encyclical entitled *On Human Labor*. The people sensed his support and their Polishness. Trade unions proliferated in schools, factories, hospitals, and shipyards as a way to influence pay and benefits. The Solidarity movement was under way.

The Polish State: Structural

Economy   Polish workers measure their prosperity by their consumption of meat [6]. With that standard it should not be difficult for the Polish government to meet the needs of its citizenry, especially since Poland is the *breadbasket of Europe*. Yet the Communist practice of an agricultural system of collectivization failed Poland miserably. The Soviet Union was at its lowest economic ebb as it entered the 1980s: zero economic growth; population increase of only 1 percent; an arms race that consumed one-sixth the GDP; inefficiency of command economy due to a waste of resources, manpower, shortages of goods, lack of parts, and shortened workdays. This reverberated throughout the Soviet sphere of influence. In

Warsaw, pro-Soviet apparatchiks were interfering deliberately with Polish economic programs to keep Poland dependent on the Soviet Union [10].

Polish economists advised decentralization, more independence for large enterprises, cost accounting, a shift from raw material production (specifically coal, which was Poland's chief source of energy) in an effort to develop other industry, such as chemical and small-scale private enterprises. But Party First Secretary Gierek would not listen, so the waste and inefficiency continued while the Polish debt to Western countries rose. By the mid-1970s, about 60 percent of Polish exports went to pay the interest on the foreign debts [10]. This meant that Poles could buy less at home, especially Polish food staples such as meat.

Starting with Gierek's economic plan, which was riddled with flaws, the Poles reacted with social eruptions that were either repressed or accommodated, and sometimes both, by the government. Although Gierek was supported by the workers, he worked at appealing to the intelligentsia and to the church to provide more effective patriotic cooperation that offered greater tolerance. His *Polish Great Leap Forward* was a broad strategy with a five-year plan (1971–1976) that concentrated heavily on industries in which export prospects were poor as opposed to modernizing private agriculture [6]. Hence, he did not win over academics or leaders in the Catholic church.

In fact, it was the intelligentsia who supported the strikers by providing their legal defense, bridging the gap between the classes in Polish society. (This relationship between the two groups would also play a significant role in the strength of the Solidarity movement.) Academicians also countered the government by instituting an underground counterculture that focused on the history and life of the nation—the Polishness.

Political System The chief complaint held by many Poles was that the regime had divided the nation against itself [11]. Its response to the nation came in the form of promises of change, then threats of use of force, then promises to negotiate with unions, then wage buyoffs—the truth was that the regime did not have the resources to offer any alternative to the despair [12]. Under Edward Gierek there was a managerial style of governance facilitated by an alleged consensual government, with the party serving as arbiter when competing interests prevailed. His massive investment schemes that were to modernize Poland led to the establishment of working groups such as the KOR (Workers Defense Committee) which spoke on behalf of workers who did not benefit from these investments. KOR's original mission was to help by collecting money to aid workers and their families, and by disseminating information through underground bulletins aimed at educating the workers. KOR claimed the rights of freedom of speech, association, and publication as provided for by the Polish Constitution, the Helsinki Agreements of August 1975 (which safeguarded human rights), and international labor agreements. (KOR was the first organization to adopt this policy in a Soviet bloc state.) KOR and another collective, Mloda Polska (Young Poland Movement, a right-wing organization), cooperated in establishing the Free Baltic Trade Union in Gdansk. The Young Poland Movement would eventually shatter the government's grasp on keeping the workers isolated and ignorant. With assistance from the movement, leaders such as Lech Walesa and Anna Walentynowicz learned about organizing political activities, activities that were taking place throughout the country. By the time Solidarity appeared in 1980, the regime had to agree to negotiations at the national level. In doing so there was an implicit acknowledgment that the regime was facing a national movement and that it could not solve the state's socioeconomic problems on its own [12].

The Polish Communist party was led by the party leadership in Moscow and state-level administration, where it was fraught with internal tensions and contradictions and distinguished by half-measures. In Poland the story was similar: the Polish party was unable to collectivize private agriculture and subjugate the church [6]. However, the precipitant cause of the Solidarity movement was in the realm of political economy. This was due to the fact that in 1979 Poland was experiencing the first decline in national income in the history of the state [6]. Food prices were increased covertly and variety and selections of food, especially meats, were being tampered with. The best products were being sold in markets that could fetch higher prices. Appeasement never served the government well because it never produced what it promised. So when the strikes began, the government could not stop their sweeping effect across the country. The Politburo in Moscow had been observing events in Poland. Many felt that the events in Poland were typical of the Poles—they always resisted; and the Soviet party leadership was already embroiled in an intervention in Afghanistan.

The conundrum for the Polish Communist leadership and the Politburo in Moscow was that the forces of Solidarity pushing for workplace reform were also pushing the party for political reforms—this made the Soviet system appear fragile [4]. Needless to say, there were many interactions and communications between Warsaw and Moscow. However, the evidence needed to shed light on the relationship between the Kremlin and the Polish apparatchik as well as to prove what took place between Warsaw and Moscow is unreliable; meetings were had, but what was said is unknown [4].

In Poland, other political parties existed, such as the Democratic party and the United Populist party, alongside the Communist party. They had at times formed coalitions with the Polish United Workers party, but they all remained dependent on the predominant Communist party. Still, at the outset of Solidarity (August 1980) about one million Communist party members joined the union; by July 1981 an additional 300,000 left the Communist party. Ten percent of Solidarity was comprised of former Communist party members [3].

It is worth noting in this section outlining the structure of the Polish state that the church—the role of religion—in Poland was not just a political safety valve; it was a dynamic force because for the Poles, *Poland was Christ among the nations* [11]. The Catholic church served as intermediary between people and party, as safe haven and spokesman when human rights abuses were observed. It also served to speak on behalf of the party when things appeared to be getting out of hand, with the fear of bloodshed.

Despite the typical censorship of Communist governments, an underground press was developing. During the late 1970s, dissidents were able to acquire copying and printing equipment smuggled in from the West, perhaps by the CIA, which was a supporter of the Polish social movement [13]. By 1980 there were over 400 different publications as well as books speaking of workers rights, human rights, and social change.

Polish authorities acted under some constraints when harassing the rapidly growing dissident movement. Too much repression might lead to public protest and unrest; it also risked antagonizing the church. The Polish party also feared condemnation from the generous U.S. government, which had been shipping surplus food products to Poland [10]. For the Polish Communist party, these workers—this Solidarity movement—would prove to be a real threat because the movement offered an alternative solution to political leadership and party membership by confronting the Communists as the "national citizen" [1].

Military   For many in Poland, the military was held in high esteem. Throughout its history the Polish army served to fight for Polish independence. Its heroic feats in World War II and the tremendous losses suffered were felt by nearly every Polish family. Given the closeness and reverence the military had with its civilian population, it is no doubt Stalin considered the Polish military to be the most unreliable—simply, *Poles do not kill Poles*. Although they were professional soldiers, they were still closely tied to the towns and villages where they were raised. Many of the soldiers in the Warsaw Pact nations went back to their farms on weekends. (This may explain the need for numerous secret service agents and domestic spying by the state parties throughout these states.) Still, Poland's military was called upon to quash the 1968 revolution in Prague and deter strikes and riots in Poland throughout the 1970s, all of which were accomplished successfully by Poland's armed forces.

With the invasion of Afghanistan in 1979, the Soviets took a renewed look at troop deployment throughout the Warsaw Pact states. The leadership concluded that it could take reductions in forces deployed in the northern states of East Germany, Poland, and Czechoslovakia without undermining those regimes or endangering Soviet hegemony [1].

The United States was also watching Soviet troop movement, specifically the Central Intelligence Agency, which was tasked by National Security Adviser Zbigniew Brzezinski to do a hypothetical white paper on the events taking place in Poland. He was suspicions of the Gdansk discussions among Polish party leaders, Lech Walesa, and representatives of the workers. The CIA white paper was not so hypothetical because the agency had discovered that while discussions were taking place in Gdansk, there was Soviet troop movement [13]. The Reagan Administration maintained its involvement with and policy toward Poland. He and the CIA recognized the Kremlin's opinion of Poland—that it was a threat and that the Polish party within was diminished in condition. The problem for the Soviets was determining how to respond.

Solidarity and the State Response

On July 1, 1980, the Polish premier announced new price increases for meat and other basic products. The price-hike announcement sparked strikes all over Poland. These strikes were uncoordinated, so government officials settled one after another by agreeing to workers' demands of increased wages. However, in late July there was a citywide strike in the town of Lublin that included the railway workers. This was very problematic for the government. If the trains stopped, so did goods shipped to the Soviet Union, and the strikers knew that. The Lublin strike was a coordinated effort and the nature of this strike served as a model for the shipyard workers in Gdansk two weeks later, on August 14, 1980.

Gdansk had suffered a notorious strike ten years earlier (December 1970) which resulted in the deaths of some employees. The shipyard workers at Gdansk never forgot this incident. On August 7, 1980 the shipyard leadership fired a popular worker, Anna Walentynowicz, for speaking out about fraud with workers' bonus allotments. The recent announcement of price increases by the Polish premier, coupled with Anna's expulsion, served as the catalyst for a strike at the shipyard. The workers had two demands: Anna was to be reinstated and the workers were to receive increased wages to counter the increase in food prices. To quash further incidents the government quickly settled the strike. As the strikers broke away to return home, Anna Walentynowicz and Alina Pienkowska (a nurse) congratulated the strikers for their success, but asked: "What about the strikers in the other places [who are] still on strike—will this government walk all over them?" The workers at Gdansk, over 1000, considered this fact and decided to return to the yard. They refused to leave. Instead, they conducted a sit-in and served as representatives for strikers throughout Poland. Soon, delegates from other Gdansk striking factories and enterprises arrived. The workers at Gdansk appealed to these strikers to stay, to coordinate their activities with them. It was Lech Walesa who put this proposal to the workers and they agreed to conduct a sit-in strike.

This coordinated effort and act of solidarity with other strikers laid the foundation for an Inter-factory Strike Committee (MKS). The committee gathered in more representatives and workers from plants, factories, shipyards, and mines throughout country. The leaders set up the Provisional Coordinating Commission to coordinate action all over the country. Soon representatives from academe, many of whom who had never been called upon by the government, were anxious to offer assistance. They joined some Gdansk intellectuals to form a group of advisers to represent the solidarity of workers throughout Poland.

The government reacted to these developments by cutting communications; telephone, rail, road, and air communications between Gdansk and the rest of the country were blocked. The party then proposed negotiations. The strikers were unwilling; they demanded that communications be restored first. In a further show of solidarity, a Catholic mass was celebrated (on August 19) during the strike outside the gate to the shipyard. This proved embarrassing for the party, as it became apparent that they lacked the ability to control the events surrounding this sit-in. Soon an agreement was reached so that the plenary meetings between government leaders and MKS leaders began. These exchanges were broadcast over the shipyard loudspeakers so that all strikers could hear what was being negotiated. Representatives for the workers called for a change in the *system of rule* as well as the *system of ownership* [6].

The strike ended on August 31 with the signing of the **Gdansk Agreement**, composed of 21 points. Some its key points were:

- Establishing a free trade union
- The right to strike
- Self-management, with workers' councils to run state enterprises
- A sliding pay scale to guard against price hikes without increasing wages

- Release of political prisoners, reinstatement of workers fired after strikes, reinstatement of university students who had been expelled
- Solidarity and society to consult with government in general on economic policy
- Restriction of exports
- Improving food supplies
- Abolishing foreign currency shops (because people with foreign currency could buy goods unavailable in other shops)
- Improved working conditions
- Health services and medical supplies
- No work on Saturdays and vacation time
- Improvement in housing availability
- Restrictions on censorship
- Ensuring Solidarity and church access to the media

The Polish government made no promise to honor this agreement, and it delayed in implementing it via the Polish Supreme Court. In fact, ratification of Solidarity did not take place until November 1980, and **Rural Solidarity** (for independent farmers) was not ratified until April 1981. Still, in eighteen days the strikers blew a hole in the Leninist myth that the working class cannot see beyond its economic wants [6]. In Poland, the Communist party held to economic issues, while the people raised their own sights. Poland was on the way to institutional pluralism.

For the first few months, the Poles were elated and enjoyed free speech and free elections of leaders in all types of civil and institutional organizations. Academics were now free to select university chancellors and speak about social reform. The strikers at Gdansk celebrated the long-awaited monument to the martyrs representing the workers killed in 1970.

1981 also saw changes in the Polish party leadership. Gierek resigned in September and was replaced by a veteran bureaucrat, Stanislaw Kania. The party, however, suffered as members emigrated to Solidarity. Of the 3000 Communist party members, 1000 left and the remaining 2000 were split between supporting the status quo and demanding change toward a more democratic party. By the summer of 1981, Solidarity had 9.5 million members (from a workforce of 12.5 million).

From the Kremlin came questions about what was happening in Poland: private ownership, no language requirement (Russian), organized labor unions, party membership at the lower levels siding with Solidarity. It was clear to Moscow that the Polish leadership had succumbed and was unable or unwilling to conduct a crackdown on this revolution. As a result, the Soviets conducted exercises and rattled sabers to make clear to the Polish party that the Moscow Politburo had a definite interest in what was happening in Poland.

Notwithstanding the Soviet presence, in the fall of 1981 (September–October) Solidarity held a congress in Gdansk. It was here that Lech Walesa was narrowly re-elected as union chairman (he was criticized for negotiating away a general strike and for behaving like an authoritarian). The significance of this congress was the crafting of the *Solidarity Program*. The program outlined a plan to combine democratic reforms with party leadership and loyalty to Polish obligations as a member of the Warsaw Pact. This made clear that Solidarity was in no way aiming to take control of the country or the government.

As Solidarity made social and political inroads, the economic situation grew worse; some people blamed Solidarity for the shortages, and many in the party blamed the workers for insisting on a five-day workweek. The national income had shrunk over 5 percent and the hard currency debt had risen to $20 billion. This downward spiral of the economy meant changes to the welfare state would be impossible [6]. The government continued to withhold promised economic information from the union and it resisted appeasement and reform. This diminished the strength of Solidarity. By September there were local strikes across the country and new trade unions were being established. There were times when as much as 50 percent of the provinces were affected by strikes that Solidarity could not stop [11]. Protests spread to all sectors: private farmers, students, writers, doctors. This led to another change in the government with the appointment of General Jaruzelski, who had served as minister of defense. He would now serve as prime minister and head of the Polish Communist party. Solidarity flexed its influence in October when Walesa made an international appeal for help which poured in from the West with caravans of trucks loaded with goods—all over the world thousands rallied with Solidarity [11]. The Soviets looked on with disdain and called for action on the part of the Polish leadership.

It is unclear just how much Soviet pressure was placed on the General to crush Solidarity, and whether or not he placated Moscow to buy time to solve the problems internally. The Soviet central committee had sent a letter to members of the Polish party committee referencing the threat to the *revolutionary achievements* of the Polish people. Some in the Politburo even accused Jaruzelski of a dual policy of agreeing with Soviet evaluations of the situation while continuing a policy of concessions to the workers. The U.S. support of Poland (evidenced in the funding provided by both the Carter and Reagan Administrations) and the straightforward denouncement of human rights by these presidents no doubt led the Soviets to believe that the Polish leadership was being affected by the Americans.

On August 14, 1981, Jaruzelski met with Brezhnev and other members of the Soviet Politburo in the Crimea. It is believed that Brezhnev expressed a very negative view of the Polish situation and insisted that force be used to crush Solidarity. To show how seriously he was viewing this, Brezhnev announced drastic reductions in oil and cotton supplies for the following year. (Further reductions came as a result of the Polish–Soviet economic negotiations that took place during the Solidarity congress.) Brezhnev also prepared for more saber rattling with a series of military exercises to take place in Ukraine, Belorussia, the Baltic Republics, and on the Baltic Sea. The Polish leadership was compelled to act on Soviet demands for a use of force.

Jaruzelski agreed to meet with Solidarity for economic talks between Solidarity and the government. In November 1981, Solidarity–church–government talks took place in Warsaw. (It was agreed by all that the church be included to serve as an honest broker as well as to provide a neutral venue.) Jaruzelski proposed the creation of a National Front in which the Communist party would be the leading force. Solidarity leaders balked. Instead, Walesa and Cardinal Jozef Glemp insisted on the establishment of a real partnership between the state and Solidarity—this time Jaruzelski balked.

In December the Solidarity executive committee convened. Some spoke radically of overthrowing the government, while others wanted free elections to local government bodies in elections due to be held in February 1982. It was finally resolved that Solidarity would agree to the establishment of a **Front of National Understanding** on the conditions that (1) a decree on trade unions would include Solidarity proposals, (2) the government would give up the provisional economic regulations, and (3) Solidarity would be guaranteed access to radio and television.

Within a few days of that meeting, Jaruzelski had a telephone conversation with Brezhnev. The Soviet leader made known that the *counterrevolution now concerns us all*. Summarily, Jaruzelski told Brezhnev there was little hope that a painful decision could be avoided, but if it was to be taken, the Poles wanted to be sure that they would carry out the action by themselves, and that the Soviet government would rescind its announcement of an economic blockade of Poland. He emphasized the severe consequences that would result from a lack of Soviet economic help.

At 6 A.M. Sunday, December 13, General Jaruzelski appeared on Polish TV and announced a state of national emergency which called for the imposition of martial law. In his speech the General told the Polish people that the "Fatherland finds itself at the edge of an abyss. ... In this situation, inaction would be a crime against the nation. One must say, enough. ... Calling on the army can have, and has, only a provisional, extraordinary aspect ... ." His speech had two appeals for popular support: (1) he promised no return to the erroneous methods and practices of the past, that the sharks of the economic underground would be exposed, and that Poland would be purged of evil; and (2) he called on Polish patriotism, appealing to the myth of the patriotic army (soldiers have clean hands). No longer should it be said that Poland stands by anarchy, an appeal that dressed up the takeover in a patriotic uniform [6]. Solidarity saw this as an attempt to liquidate the union and called for an immediate strike throughout the country. Students and academics were the first to respond, but the general strike was eventually broken in a shattering defeat. The Catholic church was told (by a liberal member of the party) that if the Poles did not conduct martial law, the Russians would. The church believed Jaruzelski and considered him the choice of a lesser evil—the church sought to prevent bloodshed.

The official justification for imposing martial law was that Solidarity leaders were preparing to overthrow the party–government leadership and abolish socialism in Poland [14]. This was not the case. Solidarity clearly stated that its demands were intended to threaten neither the foundations of the socialist regime in the country nor its position in international relations [1]. Solidarity wanted to democratize Polish socialism. Jaruzelski himself claims that he decided to impose martial law to avert a greater evil: that is, a Warsaw Pact invasion that would have led to great bloodshed. Russian Politburo documents seem to indicate that the Soviet leadership did not want military intervention in Poland, but the record is not clear [14]. Some members of the Soviet Politburo said that the Polish leadership did not want military intervention, while others claimed that they actually asked for it.

In the end the military succeeded in a coup that surprised Solidarity and left it to fragmentize and radicalize as more and more goods became unavailable. The implementation of martial law lasted from December 1981 to July 1983. Detailed preparations had been made for months, but Jaruzelski claims that he always saw martial law as the last resort and tried his best to avoid it. (Worth noting, however, is the fact that the General had at his disposal 70,000 soldiers, 30,000 Ministry of Internal Affairs functionaries, 40,000 reservists, 1750 tanks, 1900 armored vehicles, and 9000 cars [3].) Jaruzelski offered an internal solution to prevent any test of Soviet restraint and to restore conformity to the Soviet bloc at a low cost [1]. Yet, some 5000 Solidarity leaders were arrested all over the country, most in Gdansk, and were interned in special camps.

All telephone, telegraph, telex, road, and rail communications except that of the army and riot police were cut so as to prevent future coordinated national strikes.

Laying a Foundation for the Collapse of Soviet Communism

Solidarity began as a trade union movement that evolved into a movement for democratic socialism. It was a Polish national movement because it sought to democratize the Communist system by instituting self-government at all levels of economic and public life. It was a religious movement in that 98 percent of the population viewed it as a Catholic renaissance, a Catholic national identity, a Catholic respect for human and civil rights. Solidarity pioneered a new type of politics, one that included social self-organization and negotiation, and it filled the gap between the family and the nation by being a secular organization with which all could identify and in which all could have a say [1]. The events in Poland had wide ramifications. First, the Soviets learned the lesson that a planned system cannot avoid the laws of supply and demand; on the other hand, the Poles learned the economic and political price of reform; in addition, the Soviets learned that economic difficulties in the West brought harm to them [4].

Pope John Paul II believed that the history of Poland—repeatedly condemned to death—proved that it has survived and preserved its identity not because of its physical strength, but exclusively because of its own culture [1]. His first visit, in 1979, was as a world leader. He redefined power in unexpected terms and took on a national regime and an international system of government [2]. John Paul II supported Solidarity by inflicting harmful blows to Soviet Communism and by sharing his view that the Soviet Union was a hybrid system of structures forced upon a large number of ethnic groups and diverse nations.

The Solidarity movement of 1980 operated under the misperception of two key facts: the strength of Communism and the opportunity for democracy [12]. It did, however, successfully expose the weakness of Soviet Communism. Unfortunately, Solidarity was unprepared for the quickly expanded role it was expected to take. Communism had been imposed on the Poles by a foreign power which itself did not understand Communism. The Soviet view of Communism became incompatible with human nature.

Solidarity was put down through martial law in December 1981, but its ideals lived on. The fact that Poland could not be normalized (Jaruzelski spent six years attempting to return the country to Communist abnormality) is the lasting achievement of Solidarity because millions of Poles educated a younger generation about democracy—no other Communist country had done that [6]. Unlike their sister states in the Warsaw Pact, the Poles had to wait eight years longer for the collapse of Soviet Communism and the reemergence of a truly independent Poland.

Outside Poland, Solidarity paved the way for the Czech dissident movement, *Charter 77*. Hungarian dissidents were also inspired by Solidarity; they celebrated their freedom with a funeral in honor of their revolutionary leader, Imre Nagy, 31 years after his death. Even Soviet intellectuals showed interest in Solidarity, and in 1989, Soviet workers began making some of the same demands that Solidarity had made and obtained back in August 1980. Of course, the policies of *glasnost* and *perestroika* as endorsed by Mikhail S. Gorbachev played an important role in the collapse of Communism and Soviet domination over Eastern Europe. He came to save the Soviet empire; instead, he presided over its disintegration. Gorbachev's attitude was permissive compared to that of his predecessors, but this was probably due to his want of an alternative solution. However, the peoples of the Warsaw Pact countries wanted independence and democracy long before Gorbachev entered the Kremlin. Interestingly, in 1989 there were twenty-three states that defined themselves as Communist; in 2007 there were five (People's Republic of China, Cuba, Laos, North Korea, and Vietnam). Solidarity was a massive social movement that served as a civil crusade for national regeneration [1]. Solidarity also did what no army could do—it broke apart the Warsaw Pact. In this struggle, the Poles carried the banner longest with the most sustained popular push in the history of Communist Europe—Solidarity [1].

MEASURING AGENTS AND ENVIRONMENTS: STIMULI AND ACTIONS

Let us now extract elements from the narrative above as components of the environment and stimuli to facilitate analyzing the agents and their actions (behaviors). The goal is to capture the actions (reactions) of actors to stimuli and to capture the relationships between and among the actors so that an agent-based model may be developed following the methodology of Chapter 4. Recall from that chapter that the first step in agent modeling is to identify the purpose of the model. In this case the purpose is to address the research question posed earlier.

Second, we must identify the entities to be modeled and the type of output needed to answer the research question. From the account above, several entities and their relationships were identified. Those identified in Table 7-1 are used here in construction of the model. They represent the

TABLE 7-1   Model Entities and Their Relationships

Entity	Relationship
Polish government	Polish people [−M, −E], Solidarity [−P, −E]
Polish people	Solidarity [+S]
Solidarity	Polish government [−P, −M], Polish people [+E]
Catholic Church	Polish government [−P], Polish people [+S], Solidarity [+S]
U.S. government	Polish government [−P, −M], Polish people [+E], Solidarity [+P]
Soviet government	Polish government [+E, +M, +P]

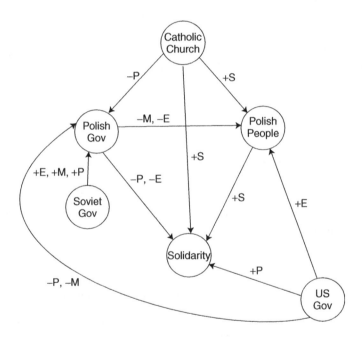

Figure 7-1   Solidarity social network.

key actors that played a role in this global event. A social network can be constructed from Table 7-1 as shown in Figure 7-1.

An explanation of the social network notation is in order. Recall that our model must represent the struggle that took place among these groups. One way to measure the outcome of this struggle is to quantify the relative power that exists among them. Power from a social sense is the ability of an actor to influence or control the choices of other actors. In international relations power relates to the total resources of an actor and is the result of using political capital in social interactions. For this case study, power is calculated as the product of capital and ability. Ability is necessary since an actor may possess significant capital, but with no means to use it, its relative power is minimized.

For this model, four power relationships were identified: *political* (P), *social* (S), *economic* (E), and *military* (M). The notation +P indicates a positive political power relationship from the originating agent to the receiving agent as depicted by the directional arrows on the social network. A −P indicates a negative relationship. The same is true for the other power variables. As an example from Figure 7-1, the U.S. government asserts a positive economic influence on the Polish people by supplying them with periodic financial aid. This link adds to the overall relative power possessed by the Polish people. Tables 7-2 to 7-5 provide the initial relative power values for all the agents in the model. Recall that these are relative values with no specific units of measure. The important point is the relative relationship of each power value to the others. These initial values were chosen based on subject-matter experts' opinions. Other methods may be used to perform a qualitative-to-quantitative mapping of relative power values. For the purpose of this model, all ability will be

TABLE 7-2   Initial Political Power Values for Each Agent

Agent	Political Capital	Political Ability	Political Power
Solidarity	5	1.0	5
Polish government	50	1.0	10
Polish people	0	1.0	0

TABLE 7-3   Initial Economic Power Values for Each Agent

Agent	Economic Capital	Economic Ability	Economic Power
Solidarity	50	1.0	50
Polish government	20	1.0	20
Polish people	10	1.0	10

TABLE 7-4   Initial Social Power Values for Each Agent

Agent	Social Capital	Social Ability	Social Power
Solidarity	5	1.0	5
Polish government	0	1.0	0
Polish people	5	1.0	5

TABLE 7-5   Initial Military Power Values for Each Agent

Agent	Military Capital	Military Ability	Military Power
Solidarity	0	1.0	0
Polish government	20	1.0	20
Polish people	0	1.0	0

assumed to be 1.0. The U.S. government, Soviet government, and Catholic church agents are considered to have unlimited capital based on their size relative to that of the other agents.

Now that we have defined the agent relationships, we can move to the next step of model construction, defining and building the agent logic structure.

MODELING HUMAN BEHAVIOR WITH AGENTS

For the agent logic model, a rule-based approach will be used. The rules will govern how each agent expends the four categories of capital necessary to achieve its goal of control over the agents against which it is competing. For this case study it is the rise in influence of the Solidarity movement over the Polish government, given the interplay of the other actors that influence the outcome.

Agents have goals that they are trying to achieve to cause their relative power to dominate other agents. The goals and their priorities for this model are delineated in Table 7-6. One can interpret this table as follows. For the Polish people agent, the top priority is to raise the Polish people's economic power above the Polish government's economic power. Its second goal priority is to have Solidarity's social power greater than the Polish government's social power. The other agents' goals can be similarly interpreted. Here, T represents *total power*.

The final necessary component in the agent design is to define the rules the agents will use to try to achieve their goals. These rules will govern how agents expend their capital in an effort to maximize their relative power. These rules are contained in Table 7-7 for every time step of the simulation

The implementation of the Soviet government rule set is shown in the logic flow diagram of Figure 7-2. This figure represents the implementation of Soviet government agent behavior in flowchart format. The diamond-shaped block represents a decision point for the agent. The rectangular blocks are actions that the agent carries out. Similar diagrams

TABLE 7-6   Agent Goals and Goal Priorities

Agent	Goal	Priority
Solidarity	Polish people [E] > Polish government [E]	1
Polish government	Polish people [E] = 0	1
Soviet government	Polish government [T] > Polish people [T]	1
Polish people	Polish people [E] > Polish government [E]	1
	Solidarity [S] > Polish government [S]	2

TABLE 7-7   Agent Rule Set

Agent	Rule
Solidarity	Provide 10% of its available economic capital to the Polish people to achieve its goal.
U.S. government	Provide a constant 3 units of economic capital to the Polish people.
	Provide a constant 5 units of political capital to Solidarity.
	Expend a constant 1 unit of political and military capital to remove the same from the Polish government.
Soviet government	Provide a constant 5 units of economic and political capital to the Polish government.
	Provide 5.5 units of military capital to the Polish government. If the Polish people's total power exceeds the Polish government's total power, provide 6.5 units of military capital.
Catholic church	Provide 1 unit of social capital to the Polish people.
	Provide 5 units of social capital to Solidarity.
	Expend 3 units of political capital to remove the same from the Polish government.
Polish people	Provide 10% of its available social capital to Solidarity.
Polish government	Expend 10% of its available political and economic capital to remove the same from Solidarity.
	Expend 10% of its available military and economic capital to remove the same from the Polish people.

can be constructed for the other agents. A complete set of these diagrams can be found at the book's Web site at http://msim.vmasc.odu.edu/global_events.

The simulation is played out over time with each of these rules applied at every time step in the simulation. In this case study, one time step is equivalent to one month of real time. Even though this is a small set of rules, trying to foresee how the agents' power changes over time is difficult for a human being to follow. From the social network of Figure 7-1 there

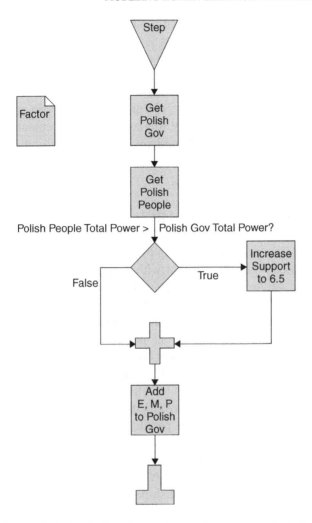

Figure 7-2 Logic flow diagram for Soviet government agent.

are many feedback connections that influence each node in the system. Having a simulation capable of automating these computations provides a ready tool for observing the behavior of the system. The behavior, when played out, may produce a result that is counterintuitive to what was expected. This is due to the generation of complex behavior even though only a small set of rules were used. This again provides the modeler with insight about the event that could not be obtained through traditional analysis methods.

These rules were derived from the case study analysis above. Although it may not be feasible to ascertain exact numbers for the parameters of the simulation, keep in mind that when modeling global events that contain social systems, one of the objectives is to get a *sense* of how the system behaves to further *understand* its behavior. Absolute predictions may not be possible.

Many different agent-based simulation packages may be used to implement this model, including Soar [15] and Repast Simphony [16]. The modeling package and the model of this case study implemented in Repast Simphony are provided on the accompanying compact disc. Let's examine the results of this model.

RESPONDING TO THE RESEARCH QUESTION

The research question called for using the Polish Solidarity movement as a case study to model human behavior with agents in an effort to capture (1) both the action and/or reaction of agents to stimuli from the environment, and (2) the relationships between and among the agents to provide a multifaceted examination of the social revolution that led to the collapse of Communism in Poland. The goal is to develop a human behavior, agent-based model that characterizes and communicates the complexity of the Solidarity movement. The following figures provide a description of this behavior played out over an 18-month period representing the duration of the main Solidarity movement. Figure 7-3 represents the rise of economic power by the Polish people over the 18-month duration of the movement. Since the people provided the labor force that produced the goods and services, how that labor force behaved had a direct bearing on the domestic product of the country. Notice that the figure shows that the relative economic power of the Polish people exceeded that of the Polish

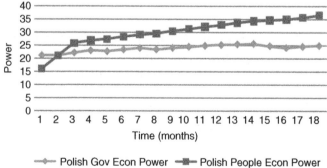

Figure 7-3   Relative economic power of the Polish people versus the Polish government.

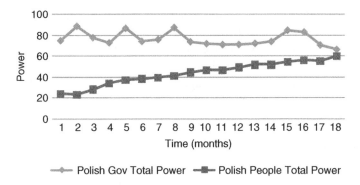

Figure 7-4   Relative total power of the Polish people versus the Polish government.

government, indicating that the people could clearly influence economic production.

It is also helpful to view the total power of each of these groups. The total power is the sum of the four power areas considered in this case study. Figure 7-4 provides a graph of total power of the people and their government. Notice that the Polish people's total power approaches that of the government near the 17-month point of the movement. When this became apparent, the Soviet Union began putting pressure on the Polish government to get control of the situation. This led to the imposition of martial law to quash the Solidarity movement while the Polish government still had a total power advantage, due primarily to its military and police force capability.

CONCLUSIONS

In this chapter we introduced the personalities and incidents inside the Republic of Poland and the founding of the Solidarity movement in 1980. The analysis explained succinctly how and why the movement took root. Included was a multifaceted look at human behavior relative to the complexity of diplomatic relationships, economics, the role of the military, the responsibility of the Catholic church, and the human capacity for freedom. The Solidarity movement was the catalyst for the eventual collapse of Poland's Communist regime and Soviet Communism. This movement depicts the age-old story of the human struggle for self-determination and sociocultural freedom.

The global event presented in this chapter was then used to illustrate implementation of an agent-based model representing that event. It provided a means to show how the agent modeling steps are actually used to construct a model, the details necessary to translate a description of the case study into model parameters, and the type of result and insight that can be gained from such a modeling activity. The reader is encouraged to develop his or her own model using this methodology to deepen the understanding of this process. An agent-based model will also be used for the case study in Chapter 8 to further emphasize this important tool.

KEY TERMS

Cold War
circular environment
apparat
Sovietization
détente
Solidarity

Poznan riots
Polish October
Gdansk agreement
Rural Solidarity
Front of National Understanding

REFERENCES

[1] Ash TG. *The Magic Lantern: The Revolution of 1989 Witnessed in Warsaw, Budapest, Berlin, and Prague*. New York: Alfred A. Knopf, 1993.

[2] Martin M. *The Keys of This Blood: The Struggle for World Dominion Between Pope John Paul II, Mikhail Gorbachev, and the Capitalist West*. New York: Simon & Schuster, 1990.

[3] Paczkowski A, Byrne M, eds. *From Solidarity to Martial Law: The Polish Crisis of 1980–1981*. Budapest, Hungary: Central European University Press, 2007.

[4] Hough JF. *The Polish Crisis: American Policy Options*. Washington, DC: Brookings Institution, 1982.

[5] Jeffreys DS. *Defending Human Dignity: John Paul II and Political Realism*. Grand Rapids, MI: Brazos Press, 2004.

[6] Ash TG. *The Polish Revolution: Solidarity, 1980–82*. New Haven, CT: Yale University Press, 2002.

[7] Weigel G. *Witness to Hope: The Biography of Pope John Paul II, 1920–2005*. New York: Harper Perennial, 2005.

[8] Weigel G. *The Final Revolution: The Resistance Church and the Collapse of Communism*. New York: Oxford University Press, 1992.

[9] Reynolds D. *One World Divisible: A Global History Since 1945*. New York: W.W. Norton, 2000.

[10] Brzezinski Z. *Power and Principle: Memoirs of the National Security Adviser, 1977–1981*. New York: Farrar, Straus & Giroux, 1983.

[11] Craig M. *Lech Walesa and His Poland*. New York: Continuum Publishers, 1987.

[12] Castle M. *Triggering Communism's Collapse: Perceptions and Power in Poland's Transition*. Lanham, MD: Rowman & Littlefield, 2003.

[13] Gates RM. *From the Shadows: The Ultimate Insider's Story of Five Presidents and How They Won the Cold War*. New York: Simon & Schuster, 1996.

[14] Pelinka A. *Politics of the Lesser Evil: Leadership, Democracy, and Jaruzelski's Poland*. Piscataway, NJ: Transaction Publishers, 1999.

[15] Soar Technology software. http://www.soartech.com/downloads.php. Accessed Nov. 14, 2008.

[16] Repast Simphony software. http://repast.sourceforge.net/index.html. Accessed Nov. 14, 2008.

CASE STUDY BIBLIOGRAPHY

Ash, Timothy Garton. *The Magic Lantern: The Revolution of 1989 Witnessed in Warsaw, Budapest, Berlin, and Prague*. New York: Alfred A. Knopf, 1993.

Ash, Timothy Garton. *The Polish Revolution: Solidarity, 1980–82*. New Haven, CT: Yale University Press, 2002.

Brzezinski, Zbigniew. *Power and Principle: Memoirs of the National Security Adviser, 1977–1981*. New York: Farrar, Straus & Giroux, 1983.

Castle, Marjorie. *Triggering Communism's Collapse: Perceptions and Power in Poland's Transition*. Lanham, MD: Rowman & Littlefield, 2003.

Craig, Mary. *Lech Walesa and His Poland*. New York: Continuum Publishers, 1987.

Gaddis, John L. *We Now Know: Rethinking the Cold War History*. New York: Oxford University Press, 1997.

Gaddis, John L. *The Cold War: A New History*. New Haven, CT: Yale University Press, 2005.

Gates, Robert M. *From the Shadows: The Ultimate Insider's Story of Five Presidents and How They Won the Cold War*. New York: Simon & Schuster, 1996.

Hough, Jerry F. *The Polish Crisis: American Policy Options*. Washington, DC: Brookings Institution, 1982.

Jeffreys, Derek S. *Defending Human Dignity: John Paul II and Political Realism*. Grand Rapids, MI: Brazos Press, 2004.

MacEachin, Douglas J. *U.S. Intelligence and the Confrontation in Poland, 1980–1981*. University Park, PA: Penn State University Press, 2002.

Martin, Malachi. *The Keys of This Blood: The Struggle for World Dominion Between Pope John Paul II, Mikhail Gorbachev, and the Capitalist West*. New York: Simon & Schuster, 1990.

Paczkowski, Andrzej, and Malcolm Byrne, eds. *From Solidarity to Martial Law: The Polish Crisis of 1980–1981*. Budapest, Hungary: Central European University Press, 2007.

Pelinka, Anton. *Politics of the Lesser Evil: Leadership, Democracy, and Jaruzelski's Poland*. Piscataway, NJ: Transaction Publishers, 1999.

Reynolds, David. *One World Divisible: A Global History Since 1945*. New York: W.W. Norton, 2000.

Walesa, Lech. *A Way of Hope*. New York: Henry Holt, 1987.

Weigel, George. *The Final Revolution: The Resistance Church and the Collapse of Communism*. New York: Oxford University Press, 1992.

Weigel, George. *Witness to Hope: The Biography of Pope John Paul II, 1920–2005*. New York: Harper Perennial, 2005.

8 Case Study: Vietnam—Johnson's War, 1963–1965

INTRODUCTION

When French President Charles DeGaulle warned President John Kennedy of impending failure in Vietnam—"we failed and you will fail"—the American president no doubt reflected on his predecessors' Cold War ideology. Truman introduced the idea of **nation building** and the promise that democracy and economic success would serve to shield any country from Communist expansion. Eisenhower had posited a **domino theory**, which asserted that unprotected countries neighboring China and the Soviet Union would fall into Communist hands, *one after another*. French Indochina—Vietnam—was one of those unprotected countries. Coupling these principles with the words of caution from DeGaulle no doubt gave pause; nevertheless, Kennedy recognized the need to expand the U.S. presence in wartorn South Vietnam.

Kennedy believed that Asians had to fight their own battles, and he was not looking to further the West's war in Vietnam. It was the responsibility of the United States as a world power that impelled him to act. Kennedy acted incrementally. In 1961 he provided equipment and military advisers on a small scale. By 1962 the number of advisers was increased from 700 to 12,000. By the end of 1963 there were over 15,000 advisers and $500 million of American aid in South Vietnam [1]. The United States now assumed military support of the war.

When Lyndon Baines Johnson took his oath of office in November 1963, the hope of nation building in South Vietnam proved futile as the threat of Communism became more real. Johnson's presidency would

Modeling and Simulation for Analyzing Global Events, By John A. Sokolowski and Catherine M. Banks
Copyright © 2009 John Wiley & Sons, Inc.

experience the highest costs in support of this war: American soldiers, American tax dollars, and American hegemony suffered while the political and social mood of the country experienced tremendous upheaval. Vietnam suffered enormous losses of life as well as near total destruction, the result of over two decades of war. Johnson labeled North Vietnam a "little pissant country," one that would "certainly not win its sovereignty from the world's most affluent society that can surely afford to spend whatever must be spent for freedom and security" [1]. The differences between these two states were stark; however, it was the differences in leadership, purpose, commitment, and determination that resulted in victory for North Vietnam.

The Chinese have a proverb which states essentially that a *dragon outside its environment is the amusement of shrimp*. The U.S. intervention in Vietnam during the Johnson Administration appears to be a dragon outside its environment. The shrimp, Ho Chi Minh's North Vietnam and the Viet Cong, appeared bemused at the inability of the dragon to function outside its environment. This case study will review the events in Vietnam from its 1945 Declaration of Independence to the escalation of the U.S. military in 1965 to illustrate the evolution of the conflict. What started as the Vietnamese war with France to overthrow colonialism transitioned to the Vietnamese war with the United States in the pursuit of nationalism. By the close of the Johnson presidency, the United States was ambivalent and averse to its commitment to the war, uncertain of its success, and disunited in purpose. Conversely, the North Vietnamese were entirely united in commitment to the war, and determined to succeed.

In this chapter we focus on the years 1963 through 1965 of the Vietnam War, dubbed **Johnson's War**, from the social network modeling perspective of both leaders to assess how patterns of relationships developed and existed among the members in each respective leader's social system and the relationships that existed in the social network. The intent of the study is to develop a model that focuses primarily on the two leaders to express how each leader's social network affected and even determined the policy, strategy, and outcome of the war during that period. This case study will provide (1) a structured methodology to represent the social network aspects of human behavior modeling, (2) a brief history of the Vietnam War, contrasting and comparing Johnson and Ho as leaders, (3) a concise explanation of social network aspects of human behavior modeling, and (4) a response to the research question.

DEVELOPING THE RESEARCH QUESTION AND METHODOLOGY

This case study is tasked with developing a social network model to analyze the patterns of relationships that developed and existed among the members in President Johnson's and Ho Chi Minh's social system as well as the relationships that existed in the social network. Agent-based modeling will be used to facilitate a representation of the behaviors that emerged. The research question for this case study can be stated: *How can agent-based modeling in a social network structure represent the behavior of President Johnson and Ho Chi Minh and the outcome of the Vietnam conflict; and is this modeling method useful in assessing alternative behaviors with changes in the agents and the structure of the social network?*

To develop a broad qualitative analysis of this case study, an assessment of the political, military, economic, social, information, and infrastructure of Vietnam is necessary to facilitate a holistic approach to understanding the environment. That review provides supplemental data for analyzing the decision making that took place within Johnson's social network regarding the diplomatic, intelligence, military, and economic factors of his policy and strategy. The case study will then engage M&S as a tool in developing a conceptual model that represents the agents within the social network system and the outcome of their actions. The goal is to develop a model of Johnson's War which can then support a prescriptive model in assessing the alternative behavioral outcomes of the war as well as model current and future interventions that share many of the same circumstances.

With the research question in hand, the following steps are recommended for this study:

1. Research the historical context of the evolution of the Vietnam War to evaluate development of the social network structures of Johnson and Ho.
2. Explain the social network structures and their effects on the behaviors of the agents which are part of the structure, with specific focus on the key actors, Johnson and Ho.
3. Select a modeling method—in this case study it is the social network aspects of human behavior modeling.
4. Apply a social network and agent-based model to validate the findings and present the results of the study.

The discussion below is based on qualitative research from primary and secondary sources on the subject of the Vietnam War. A bibliography for this case study appears at the end of the chapter.

BACKGROUND: QUALITATIVE RESEARCH

There is a plethora of literature on the subject of the Vietnam War. It has been examined from every aspect and the analysis is still ongoing. For instance, a new biography has recently been published by French historian Pierre Brocheux entitled *Ho Chi Minh: A Biography*. For many years the William J. Duiker text on Ho (*Ho Chi Minh: A Life*, 2000) was deemed a definitive study. Now, Duiker praises Brocheux's 2007 research for tracing Ho's intellectual development and providing informed judgments on issues that have provoked debate among academics. Political theorists and military historians would also find interesting a review of Gareth Porter's research (partially funded by the Lyndon Baines Johnson Foundation). In Porter's 2005 book *Perils of Dominance: Imbalance of Power and the Road to War in Vietnam*, he argues that the Cold War consensus explanation cannot be reconciled with the documentary record on Vietnam. Porter posits that a dramatic imbalance of power between the United States and the Soviet Union emerged in 1954 (when the French left Vietnam) which shaped the policies of both superpowers toward Vietnam for the next decade. Not to be overlooked is a respected work by Stanley Karnow, *Vietnam: A History*, credited as being the first complete account of Vietnam at war. Students would be well served to approach a study of the Vietnam war from a *lessons learned* perspective as the United States continues to reflect on the policy, strategy, and failure of the Vietnam intervention in the context of the war in Iraq. The modeling component of this chapter introduces the idea of restructuring the model to reflect the lessons learned and to explore the *what if*. The following brief narrative of the Vietnam intervention was drawn from these and other sources.

When Ho Chi Minh drafted Vietnam's Declaration of Independence in September 1945 he was angry with the United States for not responding to his plea for independence from the French. Ho's declaration mirrored the American Declaration of Independence, quoting its opening line: "All men are created equal." His description of eighty years of French colonialism was compelling: oppression of the Vietnamese, deprivation of democratic liberty, and inhuman laws—all emphasizing the long war the nationalists had waged against imperialist rule. Importantly, Ho was keenly aware that the United States did not act on behalf of Vietnam. To the contrary, Truman tolerated the French resumption of their colonial power post–World War II and did not act to aid Ho, despite his own doctrine of nation building and democracy as exemplified and supported in Western Europe. Ho explicitly states that "the whole Vietnamese people, animated by a common purpose, are determined to fight to the bitter end.... Vietnam has the right to be a free and independent country..." [2]. Within the declaration, Ho explicitly expresses the will of the people: The Vietnamese are united in purpose, determined to fight, and committed to the end to win their freedom. Throughout the course of the conflict between the French and then the Americans, Ho's **People's Army** personified those words. When the French conceded to the Vietnamese in 1954, the Eisenhower Administration acted to fill the vacuum caused by French withdrawal.

Eisenhower's domino theory held to the notion that the consequence of losing Indochina to a dictatorship would be detrimental to the free world [2]. This loss would come about as Communism encroached upon the region; one by one, nations would fall. Eisenhower cited nearby Laos: That government was granted its democracy in 1954 only to fracture into a right-wing, armed regime that took assistance from Moscow and North Vietnam. In 1961, President Kennedy ordered the Seventh Fleet to the South China Sea to make a strong display of America's position on the matter, politically and militarily. Moscow desired no war in Laos and shortly thereafter endorsed the appeal to a cease fire. The threat of Communist intrusion was thwarted; Laos had become neutralized by May 1961. The situation in Vietnam, however, became the subject of renewed focus.

When Kennedy came to office the United States had been supporting Ngo Dinh Diem as president of the Republic of South Vietnam. Over time, Diem and his government faltered due to corruption and internal chaos. This became widely known in May 1963 when the South Vietnamese troops opened fire on protesting Buddhist monks who had been representing the nationalist sentiment in South Vietnam. Diem did not respond appropriately to the situation, and Kennedy could do little to affect Diem's behavior. Kennedy's subsequent decision to decrease U.S. funding to the South Vietnam government resulted in a rift between the leaders. The dissipated relations led to U.S. concurrence with the South Vietnamese generals, who ousted (and later killed) Diem via a military coup in November. The political instability in Saigon undercut U.S.–South Vietnamese war efforts against a North Vietnamese takeover. Kennedy's assassination in November dropped the Vietnamese crisis into the lap of the new president, Lyndon Baines Johnson.

The Dragon: Lyndon B. Johnson

Within the first few weeks of his inauguration, Johnson had an opportunity to remove the United States from the Vietnam crisis. The assassination of Diem called for a political settlement. After the November coup, the **National Liberation Front** (the Viet Cong supporters of North Vietnam), the Secretary General of the United Nations, and the French government called for a coalition government in Saigon and the neutralization of Vietnam. Johnson said no; he wanted a triumph, not an exit from Indochina. Johnson's motives ranged from recognizing Vietnam as one of many fronts of the Cold War, to signifying clearly U.S. resolve to defeat Communism, to accomplishing something that the Kennedy Administration did not.

Throughout 1964 the war proved lethal for the United States. The Viet Cong had reportedly taken over 40 percent of the South Vietnamese countryside, the South Vietnamese army proved ineffective, and the strategic hamlet experiment collapsed.[1] On the home front, Republicans were urging military action against Ho Chi Minh in the north while Johnson's advisers were quietly planning to increase the U.S. military in the south. By the summer of 1964, the number of military advisers had risen to over 23,000. That August the **Tonkin Gulf Incident** drew a line in the sand.[2] There would be no turning back, no opportunity for diplomatic negotiations between Johnson and the Ho Chi Minh. Johnson recounted the events of the Tonkin Gulf incident to the Congress and the American public. On the heels of that message, the U.S. Congress gave the commander in chief the legal authority via a resolution to declare war against the North Vietnamese. The **Tonkin Gulf Resolution** included the following: "... to promote the maintenance of international peace and security in Southeast Asia... Congress approves and supports the determination of

---

[1]The strategic hamlet experiment was introduced by the Diem administration in 1962 to separate the Vietcong from their supporters among the South Vietnamese. Peasants were relocated from their ancestral homes to live in compounds surrounded by barbed wire and bamboo spears. There were approximately 6000 hamlets that were poorly managed and unprotected. The Diem government failed in this program so severely that it alienated many of the people it was supposed to protect.

[2]The Tonkin Gulf incident occurred in August 1964 as a result of a North Vietnamese patrol boat torpedoing the American destroyer *Maddox*, which was on an espionage mission. No U.S. casualties occurred. The U.S. destroyer then moved in closer, within four miles. With the alleged misinterpretation of sonar data indicating North Vietnamese gunboats, the *Maddox* fired wildly in what was supposed to be a counterattack. Johnson described the situation as "dumb, stupid sailors... shooting at flying fish." Johnson exploited the incident. Just as the attack was a fabrication, so was Johnson's briefing of the incident to the U.S. Congress. He garnered their full support by not divulging the incident in its entirety.

the President . . . to take all necessary measures to repel any armed attack against forces of the United States . . ." [2].

In 1965, Johnson prepared his explanation to the American public of why the United States would remain in Vietnam, why the war must be fought, and why it was a war that must be won. In speaking to an audience at Johns Hopkins University, Johnson simply stated his position as president and commander in chief: "Why are we in South Vietnam? We are there because we have a promise to keep" [3]. The speech then outlined the national pledge to help South Vietnam defend its independence. Johnson believed that dishonoring that pledge would be an unforgivable wrong. U.S. interests in national order and balance of power in the region were also mentioned in Johnson's qualifying statement: "We will not be defeated. We will not grow tired. We will not withdraw, either openly or under the cloak of a meaningless agreement . . ." [4]. The mission was clearly articulated. Johnson had determined that the United States would not consider any options like those presented to him in November 1963. The dragon had spoken.

The Shrimp: Ho Chi Minh

Ho Chi Minh had experienced much disappointment with the West. His failed attempt at gradual independence through friendly negotiations with the French and lack of support and recognition by the United States left Ho open to criticism in his own camp. Despite these failures, Ho was able to retain the loyalty of his people, and especially, his People's Army. Ho had lived among the peasants wearing the military dress of the People's Army, running and hiding in the fields with the farmers as American bombs were dropped. Ho's leadership contrasted with Johnson's in that he was a member of the People's Army in every sense of the word, as well as their leader and decision maker. His repeated appeals to cease hostilities gave credibility to his cause and his leadership. To emphasize his desire for negotiations, Ho appointed a minister of foreign affairs who was not a member of the Communist Party. The minister was a socialist, like many members of the French government at that time [5]. Ho worked diligently to encourage talks of Vietnamese independence from France; however, those efforts got him nowhere. The French would not concede to talks, let alone independence.

When Mao defeated the Kuomintang in 1949 and assumed control of the People's Republic of China, circumstances changed in Asia. There was now a new asset on Ho's balance sheet: the victory of Chinese Communism [5]. Chinese Communism coupled with the recent death of Josef Stalin and impending international dissension resulted in an irrepressible desire for France to end the engagement in Indochina. Ho took the opportunity to remind the French that the survival of the Vietminh between late 1946 and 1950 had been due to their energy, obstinacy, ingenuity, and adaptability. In May 1954, representatives from the Soviet Union, United States, Great Britain, and France met with a military delegate of the Vietminh. The delegate presented the plan for a provisional partitioning of Vietnam. Discussion among the representatives came to a close in July with an agreed upon negotiation for a temporary partition. The partition would eventually result in reunification by referendum. Ho was the mastermind of this negotiation.

The French recognized the resiliency and acumen of Ho Chi Minh. Here was a man capable of guerrilla warfare, at the same time being a statesman able to negotiate precisely what he wanted [5]. The French withdrew from Vietnam that year. Eisenhower was compelled to fill the vacuum to maintain a democratic presence in the region. Both Kennedy and Johnson would follow Eisenhower's lead by supporting the U.S.-backed South Vietnamese government throughout their administrations. In 1961, Ho's close assistant and confidant General Vo Nguyen Giap outlined a **People's War of Liberation**. The People's War strategy was to achieve Vietnamese independence and unity with an educated, mobilized, organized, armed nation—a people's army. The strategy was to wage a long-lasting battle and to put up long-term resistance fighting for a just cause [2].

When the United States entered Vietnam under the Kennedy Administration in 1961, the goal was to advise, provide equipment, and assist the South Vietnamese. After the deaths of the Buddhist monks in May 1963, Kennedy determined to establish a cooperative government. This meant that Ambassador Henry Cabot Lodge and other U.S. officials had to agree to support a coupe to remove Diem from power in South Vietnam. Significantly, this military coup would further obligate U.S. participation in the affairs of South Vietnam. Within three weeks of the ousting of Diem, the American president was assassinated. When Johnson assumed the presidency he opted to stay the course in Vietnam as outlined by Kennedy. He asked Kennedy's Secretary of Defense, Robert McNamara, to stay on and direct the crisis. McNamara remained as an adviser to Johnson until the Tonkin Gulf incident in August 1964. From that point forward, Johnson took direct control of all decision making in Vietnam.

South Vietnam had been experiencing persistent political unrest, with one government after another falling. In the spring of 1965, a military coup led by General Nguyen Van Thieu had taken control of the government. The United States was compelled to work with Thieu because the constant flux in governance left the United States hard-pressed. Given this lack of confidence in both the political and military institutions of South Vietnam, Johnson's advisers began devising a vigorous strategy to repel defeat [6]. That new strategy, presented by General William Westmoreland and the Joint Chiefs of Staff, advocated a drastic expansion of ground forces and the adoption of an offensive strategy. The military commanders adhered to a *take the fight to the enemy* offensive. Westmoreland requested 150,000 additional combat-ready troops deployed that summer. The increase in troops would be supplemented by saturation bombing in South Vietnam and the intensification of bombing in North Vietnam. The strategy was credible with one caveat: Johnson and his circle wanted to make certain not to provoke Chinese intervention. This was so important that Johnson kept tight control over the bombing, personally approving the targets and restricting attacks to the 20th parallel [6].

Two of Johnson's close advisers opposed the plan: Undersecretary of State George Ball and Washington attorney Clark Clifford. Ball warned the President that Westmoreland's plan would lead to a protracted war with an open-ended commitment. This was overturned by Secretary of Defense McNamara, who provided an argument in favor of the Westmoreland strategy: "Holding on and playing for breaks would only defer the choice between escalation and withdrawal, perhaps until it was too late to do any good" [6]. McNamara revised Westmoreland's request by recommending gradual deployment of an additional 100,000 combat forces. By July, Johnson had made his decision setting the country on a course from which it would not alter for the next three years. And what about the South Vietnamese: How did they react to this new strategy?

The Johnson Administration all but ignored the South Vietnamese. Thieu and the Saigon government were not consulted regarding the troop escalation or the bombing. There was a conspicuous absence of communication that was perceived by the South Vietnamese Ambassador Bui Diem as *unself-conscious arrogance* [6]. With the unilateral decision making and the events of the summer of 1965 came the **Americanization of the war**. Johnson felt that the honor of the United States—and his own reputation—were at stake and that he would have to Americanize the war with power wherever necessary [1].

At home, Johnson continued to practice conspicuous absence of communication. Many in his circle advised him to place these decisions before the American people. McNamara urged a declaration of a *state of emergency* due to concerns regarding domestic political implications. In doing this, Johnson could ask Congress for an increase in taxes to fund the war. This would mobilize the nation, putting it in line for intervention without a declaration of war from Congress. Johnson declined. He cited as concerns Soviet fears and Chinese reaction to these measures. Conversely, to avoid undue excitement, Johnson misled the nation as to the significance of the steps that he was taking and the direction of the war.

At the time, Johnson's advisers saw these commitments as a cautious move between extremes: withdraw or full war. Their intent was to inflict just enough pain to compel the enemy to negotiate. However, the documents now show that the Administration exercised two miscalculations: (1) They were never fully advised or informed as to what would be *enough pain to compel the enemy*: What is enough? This lead to the gross underestimation of the enemy's determination; (2) They did not foresee the eventual cost of the war, which meant that they overestimated how much the nation was willing to pay for this war [6]. Nevertheless, the new strategy was implemented.

Throughout the summer an air-war of massive proportions was conducted; the operation was called **Rolling Thunder**. This undertaking inflicted $600 million in damages, crippling North Vietnam's industrial productivity. Still, the North Vietnamese defiantly demonstrated amazing ingenuity and relentless perseverance throughout Rolling Thunder. Over 30,000 miles of tunnels were dug to house people and move supplies. They were able to quickly mobilize and rebuild infrastructure that had been bombed out just hours earlier. The Soviet Union and China assisted by providing arms and materials that were destroyed. Weather worked in support of the North Vietnamese; the monsoon season that began in September forced the curtailment of many flights. Rolling Thunder is now criticized for not achieving its goal of *inflicting enough damage to compel the enemy to negotiate*. Instead, the operation absorbed a great deal of manpower and resources that could have been used elsewhere. Ironically, Rolling Thunder is believed to have provided the rallying cry for the North Vietnamese government.

In November 1965, Ho sent word to Johnson that he was willing to meet in the interest of peace. He stipulated that talks would be possible only if a cease fire would be applied to all war operations in North and South Vietnam and that the Geneva Accords of 1954 would be accepted.[3] The Johnson Administration rebuffed Ho's stipulation. Secretary of State Dean Rusk believed that Ho was opposed to peace. In his defense Ho plainly stated: "They say we want this war to go on. How can they say such a thing...? You have seen the sufferings which the raids have inflicted on our people. They leave us with no alternative but to fight on..." [5].

---

[3]The 1954 Geneva Accords stipulated French withdrawal and temporary arrangements for a provisional government in Vietnam until North and South could come to agreeable terms.

The war did continue through the remainder of the Johnson presidency. Between November 1963 and July 1965, U.S. involvement in Vietnam transitioned from a **limited commitment** to assisting the South Vietnamese, to an **open-ended commitment** to secure the nation. Johnson escalated American involvement in the war from 16,000 soldiers in 1963 to 550,000 in early 1968 [1]. On January 22, 1973, Lyndon Johnson died of heart disease. On January 27 the war ended, with the United States withdrawing under terms of the Paris Peace Accords, designed to preserve a temporary division. The North Vietnamese soon disregarded the treaty and invaded South Vietnam, which quickly fell without U.S. support. On April 30, 1975 the South Vietnam capital of Saigon fell to the Communist forces of North Vietnam, effectively ending the war and reuniting the country.

Ho also did not witness the Paris Peace Accords, nor did he experience the reunification of his country. He died in September 1969 at his home in Hanoi at the age of 79. He, too, suffered heart failure. In his honor the former capital city, Saigon, was renamed Ho Chi Minh City in May 1975. More than 58,000 American servicemen and women lost their lives in the war [2]. The death toll and destruction that took place in Vietnam are immeasurable.

ANALYZING THE SOCIAL NETWORK STRUCTURES

With a focus on modeling the social network systems of Johnson and Ho, it is important to review the factors that affect the nodes (or agents) in the social network structure. These nodes are what determined the policy, strategy, and outcome of this period of the war. For purposes of this case study, the agents are *Johnson, Ho*, the *political circles* (advisers), the *governments*, and the *country population* of each leader. To begin the process a general overview of the purpose, commitment, and determination of both sides of the network will facilitate a compilation of the goals that are expressed later in Table 8-1.

First, there is *purpose*. When the American public protested across the country about U.S. involvement in Vietnam, it was evident the U.S. purpose and the U.S. role in Southeast Asia were not as important to them as to the President. Johnson's argument was no longer credible; it no longer satisfied the question: *Why are we there?* Global security was becoming too costly for the American public. Within Johnson's own cabinet, dissent led to the eventual departure of his key advisers, in particular Robert McNamara, who later stated in his memoirs: "We . . . who participated in the decision of Vietnam acted according to what we thought were the principles and traditions of this nation. . . . I clearly erred by

TABLE 8-1   Social Network Structure: Lyndon Baines Johnson and Ho Chi Minh

Agent	Lines of Influence
Johnson	U.S. government [+M, +E]
Ho	N.V. government [+M, +E]
U.S. advisors	Johnson [+P, +M]
N.V. advisors	Ho [+M]
U.S. government	N.V. government [−M, −P, −E]
N.V. government	U.S. government [−M, −P, −E]
U.S. population	U.S. government [+M, +P]
N.V. population	N.V. government [+M]

not forcing . . . a knock-down, drag-out debate over loose assumptions and thin analyses underlying our military strategy in Vietnam" [2].

Johnson agonized over Under Secretary of State George Ball's dissention and his argument that the war was unwinnable: that it would be long and protracted. In a conversation between Ball and Johnson, Johnson expressed concern that countries would view Uncle Sam as a paper tiger: "Wouldn't we lose credibility breaking the word of three Presidents . . .?" Ball's reply: "The worse blow would be that the mightiest power on earth is unable to defeat a handful of guerrillas" [7]. Ball's words were prophetic, for the mightiest power did lose to a guerrilla army and the loss was the first the United States had experienced.

Johnson was surrounded by brilliant minds. He retained Kennedy's circle of advisers, once considered *the best and the brightest*. Yet, when the decisions became tougher, the core group of advisers left Johnson to make those decisions on his own. Many in that circle felt that Johnson's personality played a large role in the course of events throughout the war. He wanted to have victory where Kennedy could not, and Vietnam gave him that opportunity. He was convinced that a "pissant country" could never out-distance the United States; however, Johnson failed to recognize the significance of the war for the Vietnamese. For them this was first a war to oust colonial rule, then a civil war, and finally, a war to remove U.S. dominance. The U.S. purpose in Vietnam dissipated as body counts were broadcast nightly in evening news reports. As the U.S. economy declined, the public demanded that its government concentrate on domestic issues. Johnson finally conceded the loss of purpose for the war in his final statement to the American public on March 31, 1968.

Ho, on other hand, could proclaim victory in Vietnam. He and the Vietnamese proved unconditionally committed to the war, with a do-or-die determination to succeed. Under Ho's leadership the Vietnamese remained united in purpose. Keeping the Vietnamese focused was no simple feat for the Vietnamese leader. Ho was faced with the same questions that Johnson faced: Should this war go on? How much suffering can the people endure? How many lives will be shed? Ho had to ask those questions while looking at Vietnamese people who suffered for him and for their country.

Another key factor was *commitment*. Johnson was never labeled a *hawk*: a warmonger who arbitrarily sent U.S. troops into the jungles of Vietnam. He explicitly stated that he wanted to be cautious and careful about... "sending American boys to do the fighting for Asian boys" [8]. However, after the 1964 election, Johnson resolved publicly that the American commitment would be open-ended and unqualified. Johnson articulated a number of reasons that the United States should remain committed at an escalated level: (1) to retain global balance of power, (2) to derail the Communist expansion, (3) to lessen the chance of a major war that could take place closer to home, (4) to overwhelm the opposition so that they are convinced that the war would be too costly [8]. With these justifications and the repeated promise that the end was in sight, Johnson remained committed to fight; he complied with the requests of his generals in the field to deploy more troops.

The American public's resolve and commitment was faltering. Johnson did not have a People's Army to support his decisions, in fact, quite the contrary. Johnson's decisions, his communication with the public, and the perpetuity of this military engagement created a public mood that challenged the administration. Moreover, Johnson did not have the full support of his own staff, let alone the American public, whereas his adversary, Ho, was becoming an icon to his people and to a larger global audience. During Rolling Thunder the Vietnamese simultaneously witnessed the potency of the U.S. military and experienced the impotence of a preponderance of power in meeting objectives. Ho's military used guerrilla tactics, underground tunnels, and infiltration that gave the Vietnamese the definitive advantage in this peculiar theater of war. As American troops and American citizens were becoming demoralized, commitment waned. Johnson's commitment was soon nullified by America's opposition to the war.

Ho Chi Minh and his People's Army voiced nothing but an unending commitment from the early days of the war with the French. This was their country and they would fight for their independence until the last villager. The inability of the South Vietnamese and U.S. forces to crush Ho's army of lesser-trained and ill-equipped men, women, and even children is evidence of that resolve. As the war ensued, the North Vietnamese did not waiver from their commitment. Ho appeared to have an unending supply of willing fighters, whereas the United States was counting every casualty and arguing over the draft (required military service with soldiers drawn from the civilian population via a lottery) [2].

With the commitment came the *determination*, the will to succeed. Johnson and Ho each believed that he would succeed. Both were certain of success, but their reasons differed. The U.S. military was unquestionably superior to Ho's. Its arsenal included tons of bombs, napalm, and helicopters, and it was state of the art compared to Ho's. If Johnson could only get enough personnel and equipment into the region to get the job done in one fell swoop as his generals had requested, the war could be won. Johnson's hesitance to comply *fully* with his generals, the American public outcry, and the nearly total destruction of this underdeveloped country by the most modern nation in the world proved too imposing on the imminent victory that was to be. Prominent Americans began to speak out against the war, and within the Congress, dissent was mounting. Very strong opposition came from senior citizens, blacks, females, and Americans at the lower socioeconomic levels. Arguments to end the war came from all corners: the unacceptable loss of American life; the economic destabilization; the immoral conduct of the war; an overwhelming sense of being overcommitted and overextended. Avenues for peace as recommended by the United Nations, the Vatican, and nongovernmental organizations were not acted upon by Johnson.

Ho had resources in Moscow and Beijing, but his skill as a politician, tactician, and strategist, as well as having a military that fought unshakably on the terrain with which they were accustomed, proved more valuable then helicopters and napalm. The United States was in a theater of war where the enemy knew the terrain all too well. The United States believed that precisely calculated actions could deter and/or limit aggression without the need for an escalation of U.S. forces, but Ho's army seemed limitless. The United States had no choice but to escalate troop size. Yet, in expanding the war, the United States suffered the humiliation of miscalculating the twists and turns of the war and the global perception of a state-of-the-art military opposing a rag-tag guerrilla outfit.

Thus far we have reviewed the events of 1963 through 1965. An analysis of the social network structure has also been provided in narrative form. Given those data, a table such as Table 8-1 can be used to delineate the nodes and arcs that make up the social network model of this case study. In reviewing the table, keep in mind that in this case study we reviewed primarily the character of Johnson and Ho as leaders of the war during a selected period of time. It does not apply the same level of emphasis to the military component (strategic, operational, and tactical), nor to the domestic commotion and government of both states. This matrix is a simple representation of a very complex event. The intent is to serve as a starting point for the model. The notation used in this table is discussed in Chapter 7. It is based on the concept of relative power among the agents as introduced in that chapter. (Refer to that discussion for details that form the basis for this modeling effort.)

SOCIAL NETWORK ASPECTS OF HUMAN BEHAVIOR MODELING

In Chapter 4 we discussed agent-based modeling and social networks. Recall that social network modeling helps us to understand the connections among people, whether they are political leaders, specific groups, and/or cliques in organizations. Integral to social network modeling is analysis of or disciplined inquiry into the patterns of relationships that develop and exist among the members in the social system and the relationships that exist in the social network. These relationships are characterized by *nodes*, which represent the entities sharing some relationship, and *links*, which connect the nodes together in a share relationship. The links may be one to one, many to one, or one to many. That is, each node may have only one link associated with it, or a node may originate or receive multiple links, depending on how it relates to the other nodes in the network. Closely tied to social network modeling is agent-based modeling, which focuses on the way in which social behavior emerges from the actions of agents. This case study engages agent-based modeling in a social network structure.

The model will take form from the information provided from Table 8-1. To begin creating the model, we translate the nodes and lines of influence into a network representation. This representation is depicted in Figure 8-1. Although Figure 8-1 clearly represents how all major individuals and groups were connected during the conflict, it does not provide a means to simulate these interactions in a temporal manner since the behaviors governing each node are not yet defined. These behaviors are described in the next section.

AGENT-BASED MODEL DEVELOPMENT

Given the social network described above, we can begin construction of an agent logic structure. The first step is to define the goals of each agent. Clearly, the U.S. government wanted a military victory, so a natural goal would be to have the military power of the United States be greater than that of North Vietnam. The converse was true for North Vietnam. The United States was also looking for a political win both nationally and internationally, so this must be included from a goal standpoint. In

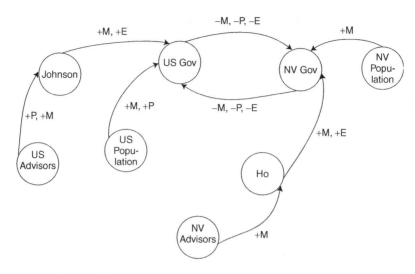

Figure 8-1   Social network representation of the Vietnam conflict.

Table 8-2 we summarize these goals in the power structure notation introduced previously.

Tables 8-3 to 8-5 provide the initial capital values for each agent. For
this simulation the U.S. and North Vietnamese populations are assumed
to have unlimited capital assets, since relative to the other agents they
are much larger. No social power is included in this model since analysis
indicated that it was not a significant factor for influencing the power
balance. As with the model developed in Chapter 7, we assume that all
relevant ability values are 1.0.

TABLE 8-2   Agent Goals

Agent	Goal	Priority
U.S. government	U.S. government [M] > N.V. government [M]	1
	U.S. government [P] > N.V. government [P]	2
N.V. government	N.V. government [M] > U.S. government [M]	1
	N.V. government [P] > U.S. government [P]	2
U.S. advisors	U.S. government [M] > N.V. government [M]	1
	U.S. government [P] > N.V. government [P]	2
Johnson	U.S. government [M] > N.V. government [M]	1
	U.S. government [P] > N.V. government [P]	2
N.V. advisors	N.V. government [M] > U.S. government [M]	1
	N.V. government [P] > U.S. government [P]	2
Ho	N.V. government [M] > U.S. government [M]	1
	N.V. government [P] > U.S. government [P]	2

TABLE 8-3 Initial Relative Economic Power Values

Agent	Capital	Ability	Power
U.S. advisors	0	0	0
Johnson	10	1.0	10
U.S. government	100	1.0	100
N.V. advisors	0	0	0
Ho	2	1.0	2
N.V. government	10	1.0	10

TABLE 8-4 Initial Relative Political Power Values

Agent	Capital	Ability	Power
U.S. advisors	10	1.0	10
Johnson	20	1.0	20
U.S. government	100	1.0	100
N.V. advisors	5	1.0	5
Ho	10	1.0	10
N.V. government	10	1.0	10

TABLE 8-5 Initial Relative Military Power Values

Agent	Capital	Ability	Power
U.S. advisors	10	1.0	10
Johnson	10	1.0	10
U.S. government	100	1.0	100
N.V. advisors	50	1.0	50
Ho	20	1.0	20
N.V. government	10	1.0	10

The remaining step in defining agent behavior is specifying an agent logic method that each agent will use for its decision making. A rule-based method (see Chapter 4) will be used for this implementation. The rules for each agent are shown in Table 8-6. The negative capital expenditure from the U.S. population toward the U.S. government represented the general lack of support by the people for the war; this had a significant effect on how the government committed resources and crafted political policy. The results of the model specified above are shown in Figure 8-2, which represents the total relative power between the two countries in this conflict. Clearly, the U.S. position eroded over time. The North Vietnamese government made great strides in keeping a superpower from achieving its goal. Their power value is shown closely approaching that of the United States, which reflects the historical outcome.

TABLE 8-6   Agent Rule Set

Agent	Rule
U.S. population	Expend a constant −5 units of political and military capital to the U.S. government.
N.V. population	Expend a constant +2 units of military capital to the N.V. government.
U.S. government	Transfer 10% of available political, military, and economic capital to reduce N.V. government power in these areas.
N.V. government	Transfer 10% of available political, military, and economic capital to reduce U.S. government power in these areas.
U.S. advisors	Transfer 10% of available political and military capital to Johnson.
N.V. advisors	Transfer 10% of available military capital to Ho.
Johnson	Transfer 10% of available military and economic capital to the U.S. government.
Ho	Transfer 10% of available military and economic capital to the N.V. government.

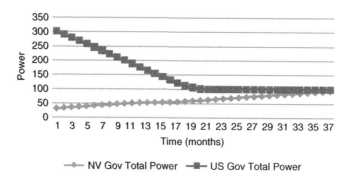

Figure 8-2   U.S. versus North Vietnamese total relative power.

The agent logic flow for this rule-based behavior is similar to that described in Chapter 7. The complete model, implemented in Repast Simphony, describes the logic flow in detail. The model can be downloaded from the book's Web site at http://msim.vmasc.odu.edu/global _events. Refer to this model to further explore alternative conditions to understand their impact on this event.

RESPONDING TO THE RESEARCH QUESTION

The research question for this case study asked: How can agent-based modeling in a social network structure represent the behavior of President

Johnson and Ho Chi Minh and the outcome during a specific period of the Vietnam conflict; and is this modeling method useful in assessing alternative behaviors with changes in the agents and the structure of the social network? A methodology was suggested that included:

1. Research the Vietnam War to evaluate development of the social network structures of Johnson and Ho.
2. Discuss these structures and their effects on the behaviors of the agents, primarily Johnson and Ho.
3. Apply social network aspects of human behavior modeling as the case study modeling method.
4. Validate the findings and present the results of the study.

The North Vietnamese fought with only success in mind. Ho did not experience the public dissent and outcry to end the war that Johnson had. Additionally, the Vietnamese continued to support Ho as the leader who rid them of colonial rule and promised victory in his fight to remove the U.S. presence, a presence that was inflicting savage acts upon them and their land. Throughout the war, even the South Vietnamese gave their support to Ho and his army. Ho epitomized Mao's theory that relatively primitive forces could prevail against more sophisticated adversaries if they practiced both *patience* and *will*. The successes of Ho's army demonstrated the irrelevancy of the type of power that the United States could project.

As in the modeling paradigm found in the case study, the modeling effort used in this Vietnam analysis showed that an agent-based model derived from the social relationships among the actors in the event could be constructed so as to characterize that event. Further, flexibility exists to explore various what-if scenarios to examine what might have happened had different decisions been made in the course of the conflict.

KEY TERMS

nation building	Tonkin Gulf Resolution
domino theory	People's War of Liberation
Johnson's War	Americanization of the war
People's Army	Rolling Thunder
National Liberation Front	limited commitment
Tonkin Gulf incident	open-ended commitment

REFERENCES

[1] Karnow S. *Vietnam: A History*. New York: Viking Press, 1983.

[2] Merrill D, Paterson TG, eds. *Major Problems in American Foreign Policy: Documents and Essays*. Boston: Houghton Mifflin, 2000.

[3] Hunt MH. *Lyndon Johnson's War*. New York: Hill & Wang, 1996.

[4] Judge EH, Langdon JW, eds. *The Cold War: A History Through Documents*. Upper Saddle River, NJ: Prentice Hall, 1999.

[5] Lacouture J. *Ho Chi Minh: A Political Biography*. New York: Random House, 1968.

[6] Herring GC. *America's Longest War: The United States and Vietnam, 1950–1975*. New York: McGraw-Hill, 2001.

[7] Barrett DM, ed. *Lyndon B. Johnson's Vietnam Papers: A Documentary Collection*. College Station, TX: Texas A&M University Press, 1997.

[8] Brown S. *The Faces of Power: United States Foreign Policy from Truman to Clinton*. New York: Columbia University Press, 1994.

CASE STUDY BIBLIOGRAPHY

Anderson, David L., and John Ernst, eds. *The War That Never Ends: New Perspectives on the Vietnam War*. Lexington, KY: University of Kentucky Press, 2007.

Averch, Harvey. *The Rhetoric of War: Language, Argument, and Policy During the Vietnam War*. Lanham, MD: University Press of America, 2002.

Barrett, David M., ed. *Lyndon B. Johnson's Vietnam Papers: A Documentary Collection*. College Station, TX: Texas A&M University Press, 1997.

Brocheux, Pierre. *Ho Chi Minh: A Biography*. Cambridge, UK: Cambridge University Press, 2007.

Brown, Seyom. *The Faces of Power: United States Foreign Policy from Truman to Clinton*. New York: Columbia University Press, 1994.

Buzzanco, Robert. *Masters of War: Military Dissent and Politics in the Vietnam Era*. New York: Cambridge University Press, 1996.

Costigliola, Frank. *France and the United States*. Boston: Twayne, 1992.

DeCaro, Peter A. *Rhetoric of Revolt: Ho Chi Minh's Discourse for Revolution*. New York: Praeger, 2003.

Duiker, William J. *Ho Chi Minh: A Life*. New York: Theia, 2000.

Gaddis, John Lewis. *Strategies of Containment*. Oxford, UK: Oxford University Press, 1982.

Haycraft, William R. *Unraveling Vietnam: How American Arms and Dipomacy Failed in Southeast Asia*. Jefferson, NC: McFarland, 2005.

Herring, George C. *America's Longest War: The United States and Vietnam, 1950–1975*. New York: McGraw-Hill, 2001.

Hunt, Michael H. *Lyndon Johnson's War*. New York: Hill & Wang, 1996.

Judge, Edward H., and John W. Langdon, eds. *The Cold War: A History Through Documents*. Upper Saddle River, NJ: Prentice Hall, 1999.

Karnow, Stanley. *Vietnam: A History*. New York: Viking Press, 1983.

Lacouture, Jean. *Ho Chi Minh: A Political Biography*. New York: Random House, 1968.

Merrill, Dennis, and Thomas G. Paterson, eds., *Major Problems in American Foreign Policy: Documents and Essays*. Boston: Houghton Mifflin, 2000.

Moise, Edwin E. *Tonkin Gulf and the Escalation of the Vietnam War*. Chapel Hill, NC: University of North Carolina, 1996.

Porter, Gareth. *Perils of Dominance: Imbalance of Power and the Road to war in Vietnam*. Berkeley, CA: University of California Press, 2005.

Vadas, Robert E. *Culture in Conflict: The Vietnam War*. Westport, CT: Greenwood Press, 2002.

9 Case Study: Cuban Missile Crisis—A National Security Emergency

INTRODUCTION

Shortly after observing the success of President Kennedy's conduct of the Cuban Missile Crisis, strategists referenced the notion of compellence, introduced by Harvard professor of economics Thomas Schelling, as a major factor in avoiding a military confrontation.[1] Schelling was an early theorist of **nuclear deterrence**, use of the threat of nuclear attack to stop an actual nuclear attack, and **compellence**, use of the threat of nuclear war to coerce an opponent into a desired action. He helped move the discussion of nuclear war from near panic (mutually assured destruction) regarding the balance of power with the Soviet Union to a more measured approach. Schelling proffered that compellence was a way to achieve a desired behavior from an antagonist state [1]. Simply stated, compellence is the bargaining power that comes from the physical harm that one nation can do to another, a *capacity for violence* that can be employed during a crisis. Unlike deterrence, which does not see conflict because it has done just what the word implies—it has deterred conflict—compellence takes place at the time of crisis, as if to say: *If you continue in this direction, you will be attacked*. The compeller takes the offensive, causing the

---

[1]Thomas C. Schelling is an American economist and professor of foreign affairs, national security, nuclear strategy, and arms control at the School of Public Policy at the University of Maryland, College Park. He was recently awarded the 2005 Nobel Prize in Economics for enhancing the understanding of conflict and cooperation through game-theoretic analysis.

Modeling and Simulation for Analyzing Global Events, By John A. Sokolowski and Catherine M. Banks
Copyright © 2009 John Wiley & Sons, Inc.

compellee to cease in her actions. In his book *The Strategy of Conflict*, Schelling explains these actions as the power to cause destruction not necessarily with traditional military strength, but with the impressive military capabilities available; notably, nuclear weapons [1]. Khrushchev had an equally impressive military capability, for the Soviet Union was also a nuclear power; yet, Khrushchev acted on the U.S. directive to remove Soviet presence and weapons from Cuba. He eventually recognized both the *lose–lose* situation and the conviction of an American president who would not shrink back from his threat.

Compellence as a strategic operative goes beyond the *threat* of destruction. It is the *poised preparedness* to attack if the correct response is not provided. Schelling asserts that compellence has to be definite. A deadline exists or tomorrow never comes. Compellence must be placed in motion to be credible. With compellence, *timing* is of the utmost importance because there can be no ambiguity. Timing must include the specifics of *when, where, what*, and *how much* [1]:

- *When—Deadlines are necessary*; with no deadline the compeller is only taking a posture. Too little time will make compliance impossible, and too much time makes compliance unnecessary.
- *Where—Stop where you are* is much clearer than *go back*, because *how far back* will be the next question.
- *What—Stop what you are doing* is not specific enough; *cease and desist in supplying missiles to Cuba* is a clear directive.
- *How much*—This question is a bit more complex because compellent threats communicate in the direction of compliance and are less likely to be self-limiting.

Compliance is always expected and it is secured with prescribed conduct that precludes a spiral of reprisals and counteractions. A compellent campaign must articulate the behavior desired and maneuver toward the compliance desired. Actions do speak louder than words. However, communicating a threat creates less uncertainty about what is being demanded. It also makes known what pressure will be kept up until demands are complied with. With compellence comes the reality of shared risk or *brinkmanship*, the competition in the risk taking. Schelling calls brinkmanship the manipulation of risk, the exploitation of danger that someone may go over the brink, dragging the other with him. As a result, compellence can set up an activity that has a potential to get out of hand. The risk of unintended disaster exists. With compellence the risk is accepted and intended—it is the potential disaster that causes the brinkmanship [1].

Compellence can also beg the question, *Who is the bad guy?* when there may be nothing but a lose–lose situation. Lawrence Freedman suggests that "compellence removes much of the relevance of distinction between aggressor and defender" [2].

The Cuban Missile Crisis pitted two nations against each other, yet it felt more like a personal challenge between two leaders who personified their nations and national ideology: U.S. President John F. Kennedy and Soviet Chairman Nikita S. Khrushchev. This case study suggests an approach to modeling national strategic decision-making during the Cuban Missile Crisis, using game theory to represent the effects of those decisions based on the concept of compellence. The exercise will use a game-theoretic perspective to model specific actions and behaviors that took place during the brief, highly dramatic national security emergency. The crisis lasted only thirteen days, but it shaped the thinking of political and military leaders for the remainder of the Cold War. The crisis depicts saber rattling at its most serious and diplomacy at its most definitive. Until the terrorist attacks of September 11, 2001, the Cuban Missile Crisis marked the nation's most panicked moment.

This chapter is presented in four subsections: (1) a concise narrative of the Cuban Missile Crisis as expressed through primary and secondary documents, (2) the concept of compellence as a strategy to obtain a behavior or action desired by a leader or state, (3) an explanation of game-theoretic modeling to analyze strategic decision making during a national security emergency, and (4) a response to the research question.

DEVELOPING THE RESEARCH QUESTION AND METHODOLOGY

The case study requires an analysis of the Cuban Missile Crisis, the second major nuclear crisis of the Cold War, and the strategy used by President Kennedy to block the Soviet Union from accomplishing its objective. The research explores the following questions: What was the relationship between the leaders of the two superpowers? What was Khrushchev's objective in arming Cuba? How did the United States respond diplomatically and militarily? Can diplomacy work without compellence?

What one can expect to learn from this case study is that diplomacy, coupled with the ability to compel, formed the national strategy necessary to avoid nuclear conflict. The research question, however, goes beyond the obvious lesson: *Using the Cuban Missile Crisis as a case study, how does game theory and the concept of compellence facilitate a visual articulation of Kennedy's strategic decision-making process and support a predictive strategy model aimed at procuring desired behaviors and avoiding military confrontation?*

Integral to this case study is a review of primary and secondary sources that provide an insight into meetings of Kennedy's Executive Committee of the National Security Council, discussions among the President's advisors, and correspondence with Nikita Khrushchev.[2] The research then engages M&S as a tool in developing a game-theoretic formulation of a conceptual model that includes the empirical data to substantiate the model. The goal is to develop a model with the predictive ability to assess possible behavioral outcomes.

With the research question in hand, a recommended methodology for this study includes four basic steps:

1. Research the literature to provide a historical context for the case study.
2. Explain the response of the actors to the use of compellence as a means to achieve a desired action.
3. Select a modeling method—in this case study it is game theory.
4. Apply game-theoretic models to validate the findings and present the results of the study.

The discussion below is based on qualitative research from primary and secondary sources on the subject of the Cuban Missile Crisis. A bibliography for this case study appears at the end of the chapter.

BACKGROUND: QUALITATIVE RESEARCH

The Cuban Missile Crisis has been examined in great detail by academics, policymakers, and military strategists. This has created a large body of literature on the topic, including many primary sources, such as *The Kennedy Tapes: Inside the White House During the Cuban Missile Crisis*, edited by

[2]Members of the Executive Committee of the National Security Council included Secretary of State Dean Rusk, Secretary of Defense Robert McNamara, Director of the Central Intelligence Agency John McCone, Secretary of the Treasury Douglas Dillon, Adviser on National Security Affairs McGeorge Bundy, Presidential Counsel Ted Sorensen, Undersecretary of State George Ball, Deputy Undersecretary of State Alexis Johnson, Chairman of the Joint Chiefs of Staff General Maxwell Taylor, Deputy Secretary of Defense Roswell Paul Gilpatric, Assistant Secretary of Defense Paul Nitze, Vice President Lyndon Johnson, Ambassador to the United Nations Adlai Stevenson, Assistant Secretary of State for Latin America Edward Martin, Special Assistant to the President Ken O'Donnell, and Deputy Director of the U.S. Information Agency Don Wilson.

October 1962

Sunday	Monday	Tuesday	Wednesday	Thursday	Friday	Saturday
	1	2	3	4	5	6
7	8	9	10	11	12	13
14	15	16	17	18	19	20
21	22	23	24	25	26	27
28	29	30	31			

Figure 9-1   Calendar highlighting the 13 days of the Cuban Missile Crisis.

Ernest May and Philip Zelikow. This book captures the discussions that were tape-recorded among President Kennedy and his circle of advisors and associates. The President and his brother Robert taped the deliberations; the editors of the book provide transcripts, commentary, and references. Another source very close to the emergency is the account written by Robert Kennedy, *Thirteen Days: A Memoir of the Cuban Missile Crisis*. U.S. Attorney General Robert Kennedy recorded the events of each day for the duration of the crisis (Figure 9-1). His book provides a behind-the-scenes story and Kennedy's assessment of the attitudes and personalities of members of the Executive Committee and the president throughout this national security emergency. These and other sources were used in the following concise narrative of the Cuban Missile Crisis.

The Policymakers

In November 1960 a newly elected president replaced a national war hero and two-term chief executive, Dwight D. Eisenhower. In a very close competition, Democrat John F. Kennedy won over Republican candidate Richard Nixon, Eisenhower's Vice President. Kennedy's inaugural address on January 20, 1961 included much of the Cold War rhetoric that the country had heard from Eisenhower. Unique to his speech, however, was Kennedy's call for sacrifice in the name of freedom and in the cause of peace. This president recognized that the world was different now because human beings held in their power the ability to abolish all forms of *human life*; yet that same power could abolish all forms of *human strife*. Kennedy's declaration to every nation that the United States would "pay any price, bear any burden, meet any hardship ... oppose any foe to assure the survival of liberty" came across as a guarantee of action.

The arms race that had begun at the close of World War II left both superpowers struggling to maintain parity. Kennedy speaks to this when he states that both nations are overburdened by the cost of these modern weapons. His inaugural speech set the tone for hope, but it also expressed a stark realism of the current state of the world. His words proved eerily prophetic, especially his petition that he would "never negotiate out of fear . . . ." Kennedy's presidency corresponded with the most dangerous phase of the Cold War, specifically the confrontations with the Soviet Union over Berlin and Cuba.

Nikita Khrushchev met John Kennedy in Vienna in June 1961. He was eager to meet the new president of the United States because his relationship with Eisenhower had soured after the U-2 affair of 1960.[3] Khrushchev was an old-style communist with peasant roots. He had learned how to survive during Stalin's harsh control of the country from 1924 through 1953. During World War II, Khrushchev served on the frontlines in the Ukraine and as head of the Communist party. His political and military experiences no doubt shaped his personality, which is said to have been impetuous and erratic. His bombast and menace made Khrushchev very much unlike Kennedy, who strove for statesmanship and sobriety.

In May 1955, Fidel Castro was given general amnesty from a prison in Cuba after being jailed for a failed coup to overthrow the government of Cuban dictator Fulgencio Batista y Zaldivar. By 1956, Castro and eighty-one followers returned to eastern Cuba in a yacht loaded with arms and ammunition. Many of these insurgents were slain or captured by Batista's forces (Castro himself was erroneously presumed dead), but the remainder found refuge in the rugged Sierra Maestra mountain range in Oriente province. For the next two years, Castro's rebels carried out a successful guerrilla campaign from their mountain stronghold, from which neither the army nor the police could dislodge them. Castro's ranks grew to several hundred as sympathizers joined the guerrilla bands. Civilian

---

[3]The incident of May 1, 1960 is known as the *U-2 affair*. It unfolded when a U.S. spy plane piloted by Francis Gary Powers was shot down 1200 miles within the border of the Soviet Union. The plane was flying from Pakistan to Norway. It is significant that this was the first time Soviet firepower could reach high-altitude aircraft (this plane could fly at an altitude of 70,000 feet). Khrushchev withheld information pertaining to the whereabouts of the plane and pilot in an effort to embarrass Eisenhower. The White House framed the incident to suggest that the pilot was flying a high-altitude weather plane but lost consciousness and accidentally violated Soviet airspace. Khrushchev told Eisenhower that his intent was not to threaten the United States but to require that the United States respect Soviet sovereignty and borders. He straightforwardly stated: "Do not fly over the Soviet Union or the socialist countries." True to his bombastic form, Khrushchev told Eisenhower: "If you don't know where our borders are, we will show you." Knowledge of the new Soviet firepower escalated the missile race: Eisenhower approved $12 billion for funding the *Atlas, Thor, Minuteman*, and *Polaris* missiles. Likewise Khrushchev increased his missile production. Khrushchev stated flippantly that the United States could take all the photographs they wanted because the Soviets were "turning out missiles like sausages from an automatic machine . . . rocket after rocket."

opposition to Batista's government increased throughout Cuba. Castro took the offensive and moved out of the mountains in the fall of 1958. Realizing that all support for his government had eroded, Batista fled with his family to the Dominican Republic in January 1959, leaving the Cuban government in the hands of the military junta. A provisional government was quickly formed with Castro as premier. The United States did not support the Cuban revolution in any way, and despite repeated overtures by Castro, President Eisenhower refused to recognize his newly formed government. In September 1960 Castro addressed the United Nations. He denounced Eisenhower for sabotaging the Cuban economy and accused the United States of imposing its will. Nikita Khrushchev was at that presentation. Without U.S. recognition and support, Castro eyed the Soviet Union. It would not be long before Cuba became a communist state. Castro boasted that his island nation would become the showcase of communism.

The superpower military competition that began in the postwar period was on a path of continued escalation. Eisenhower had explicitly outlined the nation's course of action with his policy of **massive retaliation**.[4] Khrushchev knew, however, that a change in U.S. leadership could follow with a change in policy. Khrushchev also assumed that the young, handsome, charismatic, and idealistic president who followed Eisenhower would not have the gumption that his predecessor possessed because he simply did not have the experience. The Soviet leader dismissed Kennedy as ill-equipped to deal effectively with the Soviet Union and its ambitions. Khrushchev underestimated Kennedy's resiliency.

The Crisis

On October 15, 1962, American pilots flew over Cuba with instructions to take photographs of what was taking place on the island. Six missions were flown, producing over 1000 photographs. Something was occurring—curiously, over 40,000 Soviets were in Cuba [3]. Within a few days the photos confirmed the fact that the Soviet Union was installing nuclear missiles on land just ninety miles from the U.S. mainland. Within days, the President made a decision to counter the Soviet buildup with a naval quarantine to block further deliveries of arms and an official communiqué insisting on the prompt withdrawal of the missiles that had already

[4]The postwar policy of containing communism was articulated in 1950 through NSC-68, which established a justification for U.S. military buildup acting as the world's policeman. By 1954, with Eisenhower at the helm, the United States was looking for a less expensive yet highly effective policing policy. Under the counsel of Secretary of State John F. Dulles, Eisenhower adopted the policy of massive retaliation, which shifted the emphasis from conventional military forces to nuclear deterrence as a defense strategy.

been delivered. The naval quarantine was in place by October 23. What ensued from that point on was a six-day international crisis of unprecedented severity in which the risk of nuclear war was greater than at any time before or since [4].

Given Castro's relationship with the Soviet Union and his unhappiness with the United States, the President had made clear to the Soviets that any installation of nuclear weapons capable of reaching the United States would be deemed a grave threat. In return, Kennedy had been given assurances by the Soviets that no offensive weapons would ever be installed in Cuba. This message was conveyed by Soviet Ambassador Anatoly Dobrynin in his September meetings with Robert Kennedy. President Kennedy discerned ambiguity in the assurances and was cautious. He realized that Khrushchev could present a *fait accompli* at the moment of his choice if missiles were placed on the island—and that is exactly what Khrushchev did. Still, the photographs of Soviet missiles in Cuba stunned Kennedy and his staff.

When confronted with the alarming news confirming the presence of missiles, Kennedy acknowledged that "more than words would be needed to respond" to this Soviet challenge [4]. Since Castro's coup in 1959, the United States had feared that the military dictator would turn the island into a Soviet military outpost. These fears were now a reality. Kennedy was forced to counter Soviet movement with the quarantine, which was criticized by some for overreacting and denounced by others for being too weak. The President held to the position that the United States was opposed to war and that he would maintain a policy of patience and restraint. Kennedy's caution harkens back to a theme of his inaugural address that the United States world not "prematurely or unnecessarily risk the costs of worldwide war ... but neither will the U.S. shrink from that risk at any time it must be faced" [5].

Khrushchev had been criticized at home for his inability to gain concessions with the United States regarding his plans to demilitarize Berlin.[5]

---

[5]In November 1958, Khrushchev issued an ultimatum for withdrawal of the United States from Berlin. He was speaking on behalf of the German Democratic Republic (GDR), whose government was headed by Walter Ulbricht. The East German leader claimed that all of Berlin lay in the territory of the GDR and that the Western powers had no legal basis for remaining there. Eisenhower refused to acknowledge the ultimatum to "withdraw or face war." The ultimatum had a six-month deadline. The United States called Khrushchev's bluff, causing him to be viewed as weak in the face of Western resolve. The issue of Berlin came up again with Kennedy. In June 1961, Khrushchev renewed the ultimatum and increased the Soviet arms budget. By August, Soviet forces mobilized, the borders were closed, and barbed wire and concrete blocks were set in place to begin closing off the city. A literal face-off, barrel to barrel, took place at a military

Because the West offered no concessions regarding Berlin, an increased risk of war between the superpowers was on the horizon. The Soviet Union turned its attention to Cuba as a nation in close proximity to its rival. Khrushchev recognized that Cuba was vulnerable, and he sympathized with Castro's concern regarding a U.S. invasion of the island, known as Operation Mongoose.[6] Khrushchev's concerns for Cuba's security were not, however, exactly altruistic. True, a Soviet presence in Cuba would deter a U.S. invasion; but it would also expand Khrushchev's sphere of influence—Cuba would become the first socialist country in the western hemisphere. Thus, the Soviets determined to support Castro and secure his regime. Khrushchev and his foreign minister, Georgy Kornienko, discussed the Cuban plan and how to obtain Castro's concurrence. In these talks the issue of Berlin was addressed. Soviet missiles in Cuba could leverage Soviet discussions with Kennedy regarding the situation in Berlin as well as new information from Soviet intelligence regarding U.S. missiles in Turkey. For Khrushchev, Cuba was a logical and relatively simple means of creating negotiating leverage and maintaining the balance of power.

At the time of the Cuban Missile Crisis the United States and Soviet Union presumed nuclear parity. Khrushchev is criticized for acting belligerently and willingly taking risks in Cuba. This is a significant fact for this case study because nuclear parity can make compellence problematic—nuclear compellence can lead to nuclear disaster [6]. Kennedy did not consider using nuclear weapons; instead, he relied on a preponderance of conventional forces to compel the Soviets to change their course of action. At the peak of the naval blockade, the United States had twenty-five destroyers, two cruisers, several submarines, and a large number of support ships (about 180) surrounding Cuba. Further preparations included the deployment of 250,000 conventional forces, 200 air sorties, and 90,000 marines and airborne personnel for an invasion force. On the homefront the Executive Committee was braced for an estimated 25,000 casualties [7]. (Kennedy's military strategy later proved that even in a nuclear armed environment, conventional forces are

---

checkpoint. Kennedy told Khrushchev that he must move first and continue with a gradual drawback to avoid conflict. Again Khrushchev's ultimatum had failed. On the heels of this near disaster, Kennedy made the mistake of saying: "The ratios of power today are equal." Khrushchev would not forget that boast.

[6]From January to June 1962 the U.S. Central Intelligence Agency had executed a plan, named Operation Mongoose, to remove Castro's regime. The United States committed $50 million and conducted over 6000 acts of sabotage in renewed attempts to help Cubans to take back their island. Castro called upon the Soviets to help him counter this American-born insurgency.

still necessary and can get the job done.) Khrushchev, on the other hand, had set in place all the components to arm Cuba with nuclear capability: launching pads, missiles, concrete boxes, nuclear storage bunkers—all clearly defined and ready for war.

The confrontation between the two nuclear powers took place from October 16 to October 28. The Soviets ignored the quarantine and placed a submarine in the region. Soviet ships continued sailing toward Cuba. The interaction between navies was tense until the United States observed a presumed success: Soviet ships destined for Cuba stopped on the edge of the quarantine line—twenty in all. It was soon realized that these ships were tankers and their hesitance simply created a moment of false hope. With much trepidation, the President allowed the tankers to pass. On October 25 an East German passenger ship with 1500 people had reached the barrier. Out of concern for their safety—the possibility of something going wrong—Kennedy allowed this vessel to pass through the quarantine as well.

The President began to step-up pressure by increasing the number of low-level flights over Cuba from twice a day to once every two hours. Night flights were ordered to drop red flares across the island and the State Department produced an embargo list—all in an effort to apply diplomatic pressure to the limited-force strategy of the President. If the Soviets continued to refuse recognition of the naval quarantine and if they persisted in building launching pads and preparing missile sites, a military offensive would be the only measure remaining. Members of the Executive Committee concluded that Soviet persistence would force an offensive. War was inevitable. Kennedy proceeded with preparations for a crash program on civil government in Cuba for postoccupation. Then, on October 27, a U-2 reconnaissance plane was shot down by a surface-to-air missile (SAM), causing it to crash in Cuba, killing the pilot.[7] For both the Americans and the Soviets, this gave the signal that the superpowers were

---

[7]Both Kennedy and Khrushchev were at a loss for an explanation as to the shooting of the U-2. Khrushchev had given specific orders not to fire on American U-2s for fear of an escalation. Kennedy obviously did not know about this standing order; he did, however, wonder if this attack was a mistake; thus he chose to wait before retaliating. The Soviet command had been alerted that a U-2 was spotted. Lieutenant General Stepan N. Gretchko awaited further instruction, but that instruction was delayed. The situation was tense. The Cubans had already been firing on American aircraft. The Soviet command was expecting an American attack and Gretchko was concerned that the photos would be used in an attack. True, Moscow instructed no offensive attacks on the U-2s, but it did authorize the use of SAMs for self-defense. Considering all of this, Gretchko ordered firing the SAM. In the aftermath, Khrushchev did not place any formal sanctions on Gretchko. The Cubans decorated him as a war hero.

maneuvering for war. The President felt that the noose was tightening [7], and U.S. retaliation of this incident would constitute the first step toward war. The United States acted quickly to obtain the support of the Organization of American States (OAS) for a military intervention. Yet, many questions of strategy now arose. If the United States attacked Cuba, what would be the consequences in Europe? What would happen in Berlin? Would the Soviets retaliate with an attack on Turkey? Should NATO have a say?

It wasn't long before Kennedy learned that the downing of the U-2 was not ordered by Khrushchev; rather, that decision was made by a commander on the ground. No doubt this incident caused Khrushchev to realize how quickly the crisis could escalate out of control. He was directing this operation from Moscow, nearly 6000 miles away. He was also aware of the military buildup on the U.S. mainland and in the Caribbean Sea. Importantly, he recognized that the American president had to respond to this attack. Since the confirmation of the Soviet military presence in Cuba, Kennedy had been communicating with Khrushchev regarding the need to cease and desist arming the island. Repeated communiqués, letters negotiating how to terminate Soviet activity on the island, volleyed back and forth between Washington and Moscow. It was after the attack of the U-2 plane that Khrushchev instructed his ambassador to inform the Americans of the Soviet agreement to dismantle and withdraw the missiles under adequate supervision and inspection [7].

Politically and psychologically the Cuban crisis was to accomplish much with little commitment by Khrushchev, in that a fairly limited number of intermediate-range ballistic missiles would provide him with a powerful negotiating tool to pressure the United States. Placing missiles in Cuba technically involved no breaches of international law or established arrangements among the world powers; hence, the United States was left to *assume* illegal action on the part of the Soviets to allow for a U.S. response [8]. Khrushchev had intended to bargain for concessions to the unresolved issue of Berlin and the sightings of *Jupiter* missiles in Turkey using the removal of missiles in Cuba as leverage. This was quite a gamble. Knowing that the placement of missiles in Cuba would further rupture the U.S.–Soviet relationship, why did the Khrushchev risk it? What exactly were Khrushchev's objectives, and how did he set out to accomplish them? What was Kennedy's response? What was the turning point? Why did the Soviets withdraw? Evaluating the behaviors of the two leaders will suggest and/or answer these questions.

EVALUATING BEHAVIORS

Evaluating the actions and reactions of both Kennedy and Khrushchev is the key to creating a game-theoretic model of the strategic decision making that took place during the crisis. Clearly, neither leader wanted a military conflict, yet both were equipped to engage in an exchange if the crisis were to escalate. Khrushchev had a dubious relationship with Kennedy. Keep this in mind while reviewing his objectives and his perception of the American response. Kennedy quickly learned that the Soviets established missile bases deceptively while proclaiming both privately (to members of the U.S. government) and publicly that this would never happen. In fact, as late as Wednesday, October 17, Ambassador Gromyko indicated that the only Soviet assistance being furnished to Cuba was for agriculture and land development [7]. Kennedy's assessment of Khrushchev and the Soviet presence in Cuba would now be filled with angst and grave suspicion.

Khrushchev's inability to remove the Western forces from Berlin and to demilitarize the city left him with an embarrassing political atmosphere in the Soviet Union. To turn this around, he had to find a way to redeem his foreign policy program and create some leverage in dealing with the nonconceding Americans. Establishing a military base and outfitting it with nuclear capability just ninety miles off the U.S. coast could accomplish this. Khrushchev's action in Cuba was to achieve two primary objectives: upset the balance of power and provide for the defense of the only socialist state in the western hemisphere. There were additional benefits of equal significance: Arming Cuba also repaired the missile gap that Kennedy referred to at the Vienna conference, it instigated reopening the issue of Berlin, and it would receive global notice.[8] This was important for Soviet–Sino relations. Mao Zedong and Nikita Khrushchev were not in agreement on Moscow's strategy against western capitalism. Soviet missiles in Cuba demonstrated Khrushchev's role in this ideological war. Importantly, it precluded any idea that Cuba would drift toward Chinese patronage [9].

Kennedy's reaction to the Soviet buildup in Cuba was very measured. He and his advisors determined that the U.S. objective would be the removal of the missiles. His Executive Committee considered two options:

---

[8]Khrushchev made known that he resented a remark made by Kennedy at the Vienna conference which postulated that the United States had twice as many nuclear arms as the Soviet Union and could destroy them twice over. Khrushchev's response to news reporters conveyed the seriousness with which the Soviets noted the comment. Khrushchev wanted Kennedy to understand that the United States might be able to destroy the Soviet Union *two times over* but the Soviet Union was able to wipe out the United States, even if *only once*.

a conventional air strike on the missile sites or a naval quarantine on the delivery of offensive weapons. Kennedy could justify either action as a response to Khrushchev's covert effort to arm Cuba and deliberate misrepresentation of Soviet intentions. Kennedy's public demeanor was one that exuded poise and self-control; he conveyed the ability to negotiate or to combat with equal confidence. The President made clear that any retaliation was not intended as a "victory of might, but the vindication of right" [7]. Kennedy avoided military confrontation by adhering to a strategy of compellence which required precise *timing, communication*, and a measure of *brinkmanship*.

Within forty-eight hours of sighting missiles in Cuba, Kennedy and his Executive Committee had a plan of action. Kennedy was explicit in his communication with the Soviets, calling for the immediate removal of all missiles on the island. There would be no negotiation on this matter. Kennedy had to convince the Soviets that the United States had the will to conduct an offensive intervention if necessary. He proved this by quickly deploying the quarantine and mobilizing for invasion. However, these actions did not deter the Soviets. Quite the contrary, when the directive from Kennedy was made to cease preparing missile sites, Khrushchev ignored the order and the buildup speeded up. Robert Kennedy's memoirs speak of the daily communications between the President and Khrushchev [7]. In a letter dated October 22, Kennedy made clear that it would be a mistake of the Soviet government to "not correctly understand the will and determination of the U.S." The next day, Kennedy received a response from Khrushchev accusing the President of threatening the Soviet Union with the blockade—the Chairman made clear that the blockade was not going to be observed by the Soviet Union. Khrushchev labeled the quarantine an act of banditry of which the Soviets "would be forced to take measures adequate to protect our rights." To further antagonize the naval quarantine, the Soviets deployed a submarine into the region. (This turned out to be saber rattling; but in this time of high alert, the Soviets gambled with high stakes.) The Soviets ultimately left the naval quarantine unchallenged. However, making a threat that leaves something to chance is a manipulation of risk. Added to that is the fact that the Americans considered Khrushchev's unresponsive stance to the President's directive a threat. War could easily have ensued, and up until the last moment there existed no certainty as to who would yield.

On the evening of Friday, October 26, two letters from Khrushchev were received at the White House. The first was a very long and emotional letter that included a somber reflection on the ills of war. Khrushchev asked that the United States give its word that it would not invade Cuba and lift the blockade. The second letter took on a different tone with a much harder line calling for an exchange. Khrushchev wanted the United States to remove its missiles in Turkey, and in return the Soviet Union would remove its missiles in Cuba. Kennedy responded the next day, restating the essence of Khrushchev's withdrawal letter as a guarantee that the leaders understood each other. Kennedy emphasized the Soviet removal of the weapon systems from Cuba under United Nations observation and supervision. He would facilitate the Soviet withdrawal by removing the quarantine and by giving his assurance against a U.S. invasion of Cuba. To deflect any negative criticism aimed at Khrushchev, Kennedy publicly praised the chairman's statesmanlike decision to stop building bases in Cuba [5].

Khrushchev underestimated Kennedy's resolve. It is obvious that the Soviets did not anticipate a sudden military confrontation with the United States, and as a result, they made no attempt to conceal their activity. Khrushchev's communiqués showed signs of confusion as a result of miscalculating the United States. In the end, it was he who was unable to deal with the threat posed by Kennedy. Kennedy's speedy reaction to the Soviet buildup allowed no time for Khrushchev to exploit the situation. As a result, Khrushchev did not obtain the leverage he was looking to gain with regard to Berlin.

What about Castro? Castro was influenced by his own experience of attempts of invasion by Americans. He saw Operation Mongoose as proof of unrelenting American hostility. He no doubt reveled in the humiliation experienced by Kennedy regarding prior unsuccessful attempts at invading his island. He also knew that the United States could not tolerate a socialist regime so close to home. Castro's interactions with the Soviets led him to place unwarranted trust in their intelligence assessments of an American threat. With the close of the crisis, he would learn that the Soviets had passed on unreliable and disingenuous information. It was a mistake that led to his concurrence with Khrushchev's plan. Summarily, he realized that he had been a duped by the Soviets. He angrily conceded that the Soviet presence was not so much to defend Cuba as to upset the balance of power via a Soviet military base on his island. Cuba was a pawn in this very dangerous game.

Most significant to Castro as a third actor in this drama is the fact that Castro had no inkling of the agreement reached between Kennedy and Khrushchev regarding the future of Cuba's security. When he learned that Kennedy and Khrushchev had agreed to remove their missiles from each other's backyard, he was furious, decrying Khrushchev's decision as capitulation [9]. Castro had been no part of the negotiations, yet Cuba was the bargaining chip: Soviet missiles out of Cuba for American missiles out

TABLE 9-1   Initial Payoff Matrix Setup

		Soviet Union	
		Withdraw	Maintain
United States	Blockade	x_1, y_1	x_1, y_2
	Air Strike	x_2, y_1	x_2, y_2

TABLE 9-2   Payoff Matrix with Ordinal Payoff Values

		Soviet Union	
		Withdraw	Maintain
United States	Blockade	4,4	2,3
	Air Strike	3,2	1,1

of Turkey.[9] And it was Khrushchev who traded away the bargaining chip. Castro also learned, via a radio broadcast, that the Soviets made the deal premised on the American pledge not to invade Cuba. To save face and stifle the perception of Khrushchev's exploitation of Cuba, Castro chose to blame the imperialists and to resist them as aggressors. Within a short period, Khrushchev was able to calm Castro and maintain Soviet–Cuban ties.

All three of these leaders practiced **mirror imaging**, in that each man assumed that the other saw the world in the same light as he did. All three miscalculated the consequences of their actions and did not anticipate the actions of the other [9]. Given the personal experiences and ideological and cultural differences among these men, it is understandable why these miscalculations occurred. The traditional interpretation that the Cuban Missile Crisis was a nefarious act on the part of the Soviets that blindsided the Americans provides a good starting point for a simple model of the actors. Here is an action; here are the options in model form (Tables 9-1 and 9-2). In a more advanced model (Table 9-3) we consider the personal experiences, ideology, culture, and leadership (decision-making) styles to develop a model that may serve to predict behavior.

---

[9]The nuclear warheads on the *Jupiter* missiles in Turkey were returned to the United States in late April 1963, five months after Kennedy promised that they would be removed. These missiles were obsolete and not worth salvaging; they were eventually scrapped. The Soviets never demanded verification of the removal of the missiles. Still, U.S. Secretary of Defense Robert McNamara ordered that photographs be sent to Moscow.

TABLE 9-3   Payoff Matrix with Modified Ordinal Values

		Soviet Union	
		Withdraw	Maintain
United States	Blockade	3,3	2,4
	Air Strike	4,2	1,1

GAME THEORY

In Chapter 5 we discussed game theory at length as a modeling method. Recall that game theory depends on the rational actor, who in theory is perfect but in reality is not. The Cuban Missile Crisis is a good example of the importance of *perception* as an influencing factor in decision making. Often, in the midst of crisis there is a tense atmosphere that can encourage poor *judgment*. Weighty decisions can generate conflict of *opinion* internal and external to the crisis. It was a *sense* of mutual vulnerability that precluded war between the United States and the Soviets during this period. The aspect of *humanness* is apparent because the case study refers to action–reaction and offensive–defensive responses. Kennedy's application of compellence brought a harsh reality into the crisis—concrete actions and directives were put in place to ensure no ambiguity about American resolve. In the following discussion we suggest a few models to explain and visualize the interface between the actors and the vacillating mode of behavior exhibited.

Normal Form Model

Recall from Chapter 5 the normal form representation of games. This form can be applied to construct one view of the Cuban Missile Crisis. The model that will be constructed essentially portrays the **Game of Chicken** (seeking to determine which actor will steer away from danger and lose the competition) as a representation of brinkmanship, a manipulation of risk through the exploitation of danger [10]. The initial payoff matrix for this game is depicted in Table 9-1, where x represents payoff to the United States and y represents payoff to Soviet Union. This model captures the four fundamental actions and responses and potential actions and responses that prevailed during the thirteen days. A shortcoming of the model is that its simplicity eliminates the narrative (qualitative research) of the events that took place, leaving the reader with no context for these actions and responses. This model serves to provide a basic visualization of actions and options for Kennedy and Khrushchev.

Note that the payoff values have not yet been assigned in Table 9-1. This is the most difficult part of setting up this game for analysis. These values really depend on how Kennedy and Khrushchev viewed each of these options, since they were the ultimate decision makers. One can infer their views from the discussion above and use an ordinal method for assignment, which would rank the options from most to least favorable. This method does not assign actual utility values since their specific utility functions are unknown, but it will provide a means to assign some payoff values. One possible interpretation of their preferences is given in Table 9-2 where 4 represents most favorable; 3, next most favorable; 2, next least favorable; and 1, least favorable.

With these payoff values set, one can apply the techniques of Chapter 5 to seek a solution to this game. Analyzing this table for Nash equilibria shows that only one equilibrium point exists: *blockade, withdraw*. Therefore, this set of decisions would be the rational choice.

But what if the analysis of each leader's payoff values were slightly different? Suppose that their preferences were as in Table 9-3. In this interpretation two Nash equilibria exist: *blockade, maintain* and *air strike, withdraw*. As the game shows, if the leaders had placed more importance on achieving their goal of superiority a much different outcome would probably have occurred.

There are two points to note from this analysis. One, determining the payoff values to include in the game may be difficult or impossible to achieve. Often, as in this case, these values are subjective and may not be able to be derived by a third party. Not having accurate values for the normal form of a game will lead to inaccurate results. Two, the rules for a normal form game require each player to make his choice simultaneously. The U.S. choice of a blockade came before the Soviet decision to withdraw, so this form of the game really does not accurately reflect this temporal aspect. Extensive modeling can provide an alternative analysis to account for the sequence of decisions.

Extensive Model

As noted above, the Cuban Missile Crisis played out over several days, with a sequence of decisions on both sides. On Monday, October 22, President Kennedy spoke to the American people and the free world about the events taking place in Cuba. In doing so, he raised the stakes for Khrushchev. Now the world saw Soviet expansionism in the western hemisphere and could rightly expect an American response. Kennedy's speech outlined the American response. Although the complete response

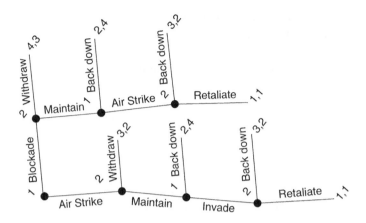

Figure 9-2 Extensive model.

was not made public, Kennedy weighed a sequence of events. One possible set of sequences is depicted in Figure 9-2 in the form of branching paths [11]. Again the payoffs here are ordinal and reflect one possible set of values by the two leaders. In this figure, Kennedy is player 1 and Khrushchev is player 2. Kennedy is faced with two principal choices: blockade or air strike. If air strike is chosen, the best outcome he could hope for is a withdrawal or backdown response by the Soviets, but not until significant U.S. forces were put in danger. If blockade is the decision, his payoff is maximized with an immediate Soviet withdrawal.

From Khrushchev's point of view, if an air strike is chosen, his best option would be to withdraw immediately because he faces loss of face if he has to back down subsequently from an invasion or risk nuclear retaliation. On the blockade option he minimizes his loss again by withdrawing the missiles immediately. If he does not do this immediately, he faces an air strike and retaliation. All these options had the potential for lower payoffs than the immediate withdrawal decision.

Enhanced Decision Modeling

A shortcoming of game-theoretic models is that they do not represent the psychological decision process that humans use to render these types of decisions. Rarely does a person quantify her utility function. If chance is involved, knowing the probabilities of all branches of a decision tree is unlikely. There may also be a very large number of branch options beyond which a person is capable of analyzing and capturing in his mind. *Enhanced decision modeling*, developed by John

A. Sokolowski, was undertaken to produce a model that would mimic the decision process of experienced decision makers at high levels [12]. The model, known as *RPDAgent* (recognition-primed decision), was designed to capture the variability inherent in the decision-making process—the humanness. The result of enhanced decision modeling is a cognitive model of decision making and a suitable means to implement the model in computational form. This model recognizes that decision making at the Kennedy or Khrushchev level seeks to best achieve the many goals that each leader had to consider. Many trade-offs must be looked at because one often uncovers goals that conflict directly with one another. As a result, decision compromises must be considered to best achieve all goals to an acceptable level. But because humans do not always behave rationally, the model accommodates personal biases as well as the unknown; a person may not be able to determine the optimum choice because not all possible outcomes and probabilities are known. This type of modeling includes an understands of how experience or situational awareness affects decisions. The Cuban Missile Crisis serves as a good case study because it was apparent that the decision makers had quite opposite experiences. Also included in RPDAgent modeling are four factors that affect behaviors: cues, goals, actions, and expectancies.

In the section "Evaluating Behaviors," these and other factors were discussed. A list comparing the experiences and situational awareness of Kennedy and Khrushchev, given in Table 9-4, begins the RPDAgent modeling process. Although it is beyond our scope in this chapter to completely describe Kennedy's and Khrushchev's decisions with RPDAgent, suffice it to say that this method goes far beyond the rational representation of these decisions via game-theoretic constructs. This is because RPDAgent modeling includes methods that capture the essence of human decision making based on life experience, goals, expectancies, and psychological characteristics that go into characterizing the human decision process. (The reader is referred to Sokolowski's dissertation for a detailed explanation of the RPDAgent model [14].)

Comments on the Model   Adding Castro to this model would add another layer of complexity to the decision making. This can be done via game theory by adding Castro as a third player and expanding the game to an *n*-person game with its associated multilevel payoff matrix. Castro could also be added as a separate RPDAgent model that captured his decision process via constructs similar to Table 9-4. The Castro agent could then interact with the Kennedy and Khrushchev agents in a multiagent system

TABLE 9-4   Experiences and Situational Awareness of Kennedy and Khrushchev

	Kennedy	Khrushchev
Experience	Supported an ideology of democratic ideals with a realist approach to the world climate Considered an intellectual and composed leader Recognizes the burden of the arms race History of poor personal and public relations with Khrushchev U.S.–Cuban relations highly strained Sought concurrence from the OAS for U.S. action against the Soviets	Old-school communist, however more sympathetic to the years of compromise the Soviets endured; practices goulash communism—putting meat into the pot for every table Considered erratic, passionate–emotional, bombastic Recognizes the burden of the arms race but is resolved to stay with it Thought Kennedy too young and inexperienced; felt sorry for Americans whose president was so deficient Looked to Castro as a socialist partner Wanted to block Cuban–Sino relations Wanted to show China tough on capitalists Missile gap–balance of power issue Berlin issue Missiles in Turkey Strained domestic relations Soviet aid to Cuba not an illegal action
Cues	United States made repeated statements regarding Soviet aid to Cuba Received word that Soviets would not arm Cuba; this deception caused suspicion Concerned about NATO and OAS	This action a departure from traditional policy Talks of Cuban defense Promises to refrain from arming Cuba were untrue

TABLE 9-4   (*Continued*)

	Kennedy	Khrushchev
Goals	Remove Soviet arms and presence in Cuba and region Minimize the chance of a military conflict	Berlin demilitarized Missiles in Turkey removed
Actions	Very measured approach to counter Soviet action (within 48 hours had a plan of action) Naval quarantine Communiqués Able to step-up pressure Unsuspected missile attack of U-2 Mobilize for war	Soviets made mistakes in the process of arming Cuba; it was done overtly with little effort to hide what was taking place Arm Cuba Continued objective of arming Cuba Unsuspected attack of U-2 caused great concern of escalation—provoked agreement with U.S. Withdrew and abandoned Cuba
Expectancies	Positive Soviet response No war Remove Soviets	Soviet presence would weaken Kennedy's resolve Balance of power to control Cuba and leverage negotiations between powers
Personal stakes	Nuclear war Cuba	Commitments, thoughts, words shape his behavior Castro

that took the cues for its decision making from the other agents in the environment.

RESPONDING TO THE RESEARCH QUESTION

Let us slightly rephrase our research question for the Cuban Missile Crisis: *How does game theory and the concept of compellence facilitate a visual articulation of Kennedy's strategic decision-making process, and can this type of modeling support a predictive strategy model aimed at procuring desired behaviors and avoiding military confrontation?* The answer came as a formal modeling methodology that introduced traditional game-theoretic models and a more sophisticated enhanced decision model that used multiagent system simulation. The methodology for this case study included four steps:

1. Research the literature to provide the historical context and the necessary primary documents.
2. Explain the reaction to the application of compellence as a means to achieve desire behavior.
3. Select a modeling method—in this case study it is game theory.
4. Apply a game-theoretic model to validate the findings and present the results of the study.

The study concluded that this methodology could be used to analyze contemplated strategies in time of crisis by developing three models, two traditional and one advanced, that probe the intentions, behaviors, and outcomes of the actors. However, the study also pointed out that determining appropriate values in one's payoff matrix could be problematic. These values typically represent individual utility that is difficult to measure from a third-person perspective. Even if there is direct access to the decision maker, asking him to characterize a set of decision values in this manner is not an exact science.

What other case studies can be analyzed using traditional game theory: case studies where actors (leaders) or states seek to answer questions of security such as what objectives (perceived) or circumstances a state deems to be a threat? What is the goal of the leader or state, and what are the options for achieving that goal? What are the costs and benefits of each option for the leader or state? What is the best option [13]? The purpose of these questions is to develop complex models that probe intentions, explain behaviors, and proffer outcomes.

Kennedy's leadership style seemed to have reflected the ideas of British strategist Basil Liddel Hart: keeping strong, keeping cool, and keeping patient. Hart warns that one should never corner an opponent and should always assist him to save face [3]. Following this approach requires being able to fully understand or *read* a crisis, including the sources of its urgency, options available, the interests of those involved, and the options available to them. Kennedy's management of the crisis supported a new method for dealing with crisis: **flexible response**.[10] In sum, his use of compellence allowed him to move forward one step at a time, increasing the pressure on Khrushchev at each stage. In doing this he was able to

---

[10]Flexible response emphasized military expansion to ensure that there would be no missile gap as well as to maintain the use of conventional and nuclear arms.

assess the *will* of the Soviets, then determine options for settlement or added pressure.

Kennedy recognized that as a politician he was obligated to lead the country through crisis and that this had to be with a show of fortitude, certainty, and toughness. He also recognized that his opponent would face the same challenges. As both chief diplomat and commander in chief, he did not view Khrushchev as a singleminded foe with whom no common ground could be shared, nor did he think communication and cooperation were impossible [3]. Khrushchev, however, made it difficult for Kennedy to live up to his philosophy: all the more reason why the outcome of the Cuban Missile Crisis is astonishing.

Modeling human behavior such as this is challenging. Rational actor models assist with visual representations of decision making, and advanced models provide a characteristic account of events that affect decision makers. Modelers of game theory should continue to develop modeling techniques that articulate and illustrate a holistic analysis of decision making by incorporating the all-important humanness component.

KEY TERMS

nuclear deterrence

compellence

massive retaliation

mirror imaging

game of Chicken

flexible response

REFERENCES

[1] Schelling C. *The Strategy of Conflict*. Cambridge, UK: Cambridge University Press, 1960.

[2] Freedman L. *The Evolution of Nuclear Strategy*. New York: St. Martin's Press, 1997.

[3] Freedman L. *Kennedy's Wars: Berlin, Cuba, Laos, and Vietnam*. New York: Oxford University Press, 2000.

[4] Bundy McG. *Danger and Survival: Choices About the Bomb in the First Fifty Years*. New York: Vintage Books, 1990.

[5] Sorensen TC., compiler. *"Let the Word Go Forth:" The Speeches, Statements, and Writings of John F. Kennedy*. New York: Doubleday, 1988.

[6] Coffey JI. *Strategic Power and National Security*. Pittsburgh, PA: University of Pittsburgh Press, 1971.

[7] Kennedy RF. *Thirteen Days: A Memoir of the Cuban Missile Crisis*. New York: W.W. Norton, 1969.

[8] Linden CA. *Khrushchev and the Soviet Leadership, 1957–1966*. Baltimore: Johns Hopkins University Press, 1966.

[9] Munton D, Welch DA. *The Cuban Missile Crisis: A Concise History*. Oxford, UK: Oxford University Press, 2006.

[10] Brams SJ. Game theory and the Cuban missile crisis. In *Plus*, Issue 13, 2001. http://plus.maths.org/. Accessed Nov. 2008.

[11] Deiwiks C. Game-theoretic models of the Cuban missile crisis. In *Game and Decision Theory*, 1987. http://www.apparently.de/sources/essays/Game_Theory_Missile_Crisis_Essay.pdf. Accessed Nov. 2008.

[12] Sokolowski JA. Enhanced decision modeling using multiagent system simulation. *Simulation* 2003;79(4):232–242.

[13] Allison GT, Zelikow P. *Essence of Decisionmaking: Explaining the Cuban Missile Crisis*. Boston: Pearson Education, 1999.

[14] Sokolowski JA. Modeling the decision process of a joint task force commander. Ph.D. dissertation, Old Dominion University, 2003.

CASE STUDY BIBLIOGRAPHY

Allison, Graham T., and Philip Zelikow. *Essence of Decisionmaking: Explaining the Cuban Missile Crisis*, Boston: Pearson Education, 1999.

Betts, Richard K. *Nuclear Blackmail and the Nuclear Balance*. Washington, DC: Brookings Institution, 1987.

Bundy, McGeorge. *Danger and Survival: Choices About the Bomb in the First Fifty Years*. New York: Vintage Books, 1990.

Coffey, J. I. *Strategic Power and National Security*. Pittsburgh, PA: University of Pittsburgh Press, 1971.

Cowen-Karp, Regina, ed. *Security with Nuclear Weapons*. Oxford, UK: Oxford University Press, 1991.

Dallek, Robert. *An Unfinished Life: John F. Kennedy, 1917–1963*. New York: Little, Brown, 2004.

Freedman, Lawrence. *The Evolution of Nuclear Strategy*. New York: St. Martin's Press, 1997.

Freedman, Lawrence. *Kennedy's Wars: Berlin, Cuba, Laos, and Vietnam*. New York: Oxford University Press, 2000.

Gutierri, Karen, et al., The integrative complexity of American decision makers in the Cuban missile crisis. *Journal of Conflict Resolution*, Dec. 1995.

Kennedy, Robert F. *Thirteen Days: A Memoir of the Cuban Missile Crisis*. New York: W.W. Norton, 1969.

Lebow, Richard, and Janice G. Stein. *When Does Deterrence Succeed and How Do We Know It?* Ottawa, Ontario, Canada: Canadian Institute for International Peace, 1990.

Linden, Carl A. *Khrushchev and the Soviet Leadership, 1957–1966*. Baltimore: Johns Hopkins University Press, 1966.

May, Ernest R., and Philip D. Zelokow, eds. *The Kennedy Tapes: Inside the White House During the Cuban Missile Crisis*. Cambridge, MA: Harvard University Press, 1997.

Munton, Don, and David A. Welch. *The Cuban Missile Crisis: A Concise History*. Oxford, UK: Oxford University Press, 2006.

Richter, James G. *Khrushchev's Double-Bind*. Baltimore: Johns Hopkins University Press, 1994.

Schelling, Thomas C. *The Strategy of Conflict*. Cambridge, UK: Cambridge University Press, 1960.

Schlesinger, Arthur M. *A 1000 Days: John F. Kennedy in the White House*. Boston: Houghton Mifflin, 1965.

Sorensen, Theodore C., compiler. *"Let the Word Go Forth:" The Speeches, Statements, and Writings of John F. Kennedy*. New York: Doubleday, 1988.

Talbott, Strobe, ed. *Khrushchev Remembers*. Boston: Little, Brown, 1971.

Taubman, William. *Khrushchev: The Man and His Era*. New York: W.W. Norton, 2004.

Ulam, Adam B. *Expansion and Co-existence: Soviet Foreign Policy 1917–1973*. New York: Praeger, 1974.

INDEX

Modeling and Simulation for Analyzing Global Events, By John A. Sokolowski and
Catherine M. Banks
Copyright © 2009 John Wiley & Sons, Inc.

Printed and bound by CPI Group (UK) Ltd, Croydon, CR0 4YY

16/04/2025

14658366-0004